GUESTS IN YOUR GARDEN

GUESTS IN YOUR GARDEN

Michele Davidson
Illustrations by Eve Corbel

ARSENAL PULP PRESS
VANCOUVER

ARSENAL PULP PRESS
103-1014 Homer Street
Vancouver, B.C.
Canada V6B 2W9
arsenalpulp.com

The publisher gratefully acknowledges the support of the Canada Council for the Arts and the B.C. Arts Council for its publishing program, and the support of the Government of Canada through the Book Publishing Industry Development Program for its publishing activities.

Book design by Lisa Eng-Lodge
Production Assistant Judy Yeung
Sidebar research by Helen Godolphin
Cover photo by C. Krebs/Tony Stone Images
Printed and bound in Canada

CANADIAN CATALOGUING IN PUBLICATION DATA:
Davidson, Michele
Guests in your garden

Includes bibliographical references.
ISBN 1-55152-097-4

1. Garden pests. 2. Invertebrate pests. I. Corbel, Eve, 1948- II. Title.
SB931.D38 2001 635'.0496 C2001-910458-8

TABLE OF CONTENTS

Acknowledgments 7
Foreword *by David Tarrant* 11
Introduction 13

THE BOLD 15
Ants 16
Bees 19
Wasps 23
Cockroaches 26
Termites 28
Beetles 31

THE BEAUTIFUL 37
Moths & Butterflies 38
Caterpillars 41
Damselflies & Dragonflies 44
Mayflies 47

THE MOTLEY CREW 49
Praying Mantids 50
Grasshoppers & Crickets 53
Walking Sticks 56
Thrips 58
Earwigs 60

THE IRRITATORS 63
Fleas 64
Flies 67
Mosquitoes 69
Craneflies 71
Fruit Flies 73

The Many-Legged 75
Centipedes 76
Millipedes 78

The No-Legged 81
Slugs 82
Snails 85
Earthworms 88
Leeches 91

Bugs: The Real Thing 93
Waterbugs 95
Aphids 97
Scale 99
Spittlebugs 101
Cicadas 102
Mealybugs & Sowbugs 104

The Not-So-Scary After All 107
Spiders 108
Scorpions 113
Mites & Ticks 115

Bibliography 117

ACKNOWLEDGMENTS

Many people deserve recognition for their contributions to this endeavour. First and foremost is illustrator extraordinaire, Eve Corbel. Her drawings capture perfectly our mutual admiration for insects. And heaps of praise for Helen Godolphin, our sidebar researcher, for her enthusiastic contribution to this book.

Thanks also to Brian Lam, Arsenal Pulp Press' publisher – my hero. His calmness in my occasional moments of panic helped pull me through. Bless you, Brian, for your example and your kindness.

Finally, I express my gratitude to my mother and father, Ron and Gail Davidson, for always knowing when to provide emotional support. And to my beloved spiritual friend, Zasep Rinpoche, for asking me to promise not to kill spiders.

For Rinpoche

FOREWORD

BY DAVID TARRANT

Michele Davidson takes us on an amazing adventure in *Guests in Your Garden*. With her sensitive insight into the world of insects, she opens our eyes to a part of nature on which we are very dependent – the insect world.

I've been gardening all my life and have come to know many of the garden guests Michele writes about so vividly. And in my thirty-two years at the University of British Columbia's Botanical Garden, I have witnessed some beautiful sights, none so moving as those created by Mother Nature and by our crawling friends: dancing clouds of tiny flies in the glow of late evening summer sunlight. Spring flocks of bush tits merrily eating the aphids on apple leaves. A pileated woodpecker just a few feet away, attacking an old tree stump for his supper of termites; if the termites weren't there, neither would the lively woodpecker. But perhaps my favourite memory is of wandering among the Asian Garden's remarkable plant specimens. The sight of spider webs delicately arranged between the dew-laden branches of the brightly hued Japanese maples was truly magical.

You may be surprised to know that I am a fan of slugs because of the good they do in our gardens. A few years back, while taping the TV program *Canadian Gardener*, I visited a nature park to do a show on slugs. This was the first opportunity I had to hold one of the Pacific Northwest's rare banana slugs. I was quite surprised by how gently it moved across the palm of my hand. That day, I learned the important role this slug played in breaking down natural leaf litter on the forest floor. I now spend time putting slugs *into* the compost pile to do their fine work, instead of trying to get rid of them.

It's too bad that many gardeners are still conditioned to believe that anything crawling should be eradicated as quickly as possible. But I am encouraged by the movement of positive change taking place lately in the gardening community. More and more gardeners are turning to organic methods as they learn how pesticides harm rather than benefit their gardens; they see that chemical quick fixes kill all the good guys too. The city of Halifax, in eastern Canada, is the only community that I know of in Canada to ban the use of pesticides and herbicides in the garden. I hope there will be a groundswell of support for similar bills across North America. And this is sure to happen when people become more educated about pesticides.

Enjoy reading Michele's fun but also educational book. I hope it leads you to discover the truly wonderful creatures that live, work, and play in your garden.

INTRODUCTION

We've all had moments or experiences that dramatically alter our perspective. I had my first one long ago at the age of six, watching a cartoon of Dr. Suess' *Horton Hears a Who* on my family's black and white television.

Horton is a lovably goofy elephant who hears a voice in a dustball. He is thunderstruck, in the endearing way only large cartoon animals can be, to discover an entire world of miniature beings within. Naturally, they became fast friends. From that moment on, my child's mind imagined a vast array of life in places like clouds, dandelions, pine cones, and closets. Sometimes it was scary, because not all the beings I "saw" were friendly, but most of the time I was happy to contemplate the creatures that might be living all around me.

I thought of Horton again, years later, when I became a practitioner of Tibetan Buddhism. For just as I embraced Horton's kindly message about the preciousness of even the smallest life, so too did I embrace the Buddha's teaching not to harm any sentient being, be it human, animal, or insect. I freely admit that sparing the lives of the creepiest of the creepy crawlies took a bit of adjustment. And it was with no small amount of terror that I trapped the impressive spiders that skittered across the floor of my basement apartment so I could release them outdoors.

Making the decision not to kill insects gave me the opportunity to observe the insect world more closely. I learned that their diversity is truly astounding – there are hundreds of thousands of insect species, each unique and endearing in its own way. In insects, I see examples of cooperation, mutually beneficial behaviour, and tolerance. Even predation, while not so lovely, is nature's way of maintaining balance.

Given my way of looking at the world, writing a book called *Guests in Your*

Garden felt logical, natural. It has deepened my appreciation that all life forms, no matter how small, have their special role to play in the ecosystem.

It seems to me that many people – gardeners included – have lost sight of the fact that nature, when left alone, maintains its equilibrium. They don't understand that we upset nature's careful balance with our use of pesticides. How ironic that in trying to free themselves from slugs, caterpillars, and aphids, gardeners destroy their soil and ultimately increase their "pest" problems.

I believe that gardeners have a vital role to play in helping to protect the web of life on Earth. We are the caretakers of many of our planet's green spaces. And we can lead the way to a healthier planet by choosing not to kill the creatures that live in our gardens simply because they are inconvenient.

I encourage you to take a look at the amazing world of insects that exists in *your* backyard. Watch carefully and you'll soon see that the ants that help open your peony blooms are dependent on aphids. And that caterpillars turn into moths so stunning you'll think they're butterflies. Even slugs and cockroaches redeem themselves by recycling your garden refuse.

You'll see too that garden chemicals kill beautiful fireflies, and the earthworms that make your soil richer. And how the busy bees, necessary for pollination, will die or go elsewhere when you upset the balance in your garden.

I hope that this book will reawaken the sense of wonder you had as a child, when you laid on your stomach in tall grass and watched the insects that lived there – long before you learned to fear or loathe them.

Welcome the small guests in your garden! Let them do the jobs that nature has set for them. And, while you're at it, put your feet up and read *Horton Hears a Who*. You'll never look at a dustball, or your garden the same way again.

THE BOLD

ANTS

ants

There are nearly 8,000 species of ants, ranging in colour from jet black to pale yellow. Ants are among the most shapely guests in our gardens – their impossibly small waists provide flexibility so the tiny creatures can waggle their bottoms to sting, lay eggs, and communicate.

Most of us think of ants as crawling creatures because of their six legs, but to my surprise, I recently learned that they can fly too. But ants don't fly often. They generally become airborne only when it's time to mate.

The relationship between ants and plants is an important one. Ants disperse seeds and in some cases help open tightly budded flowers. Have you ever seen ants in your peony blooms? They're helping to pry open the massive buds. Plants provide nutrition for ants and, in return, ant poop provides nutrients for plant and seed growth. Ants are also active recyclers. They're second only to earthworms in the amount of nutrients they return to our garden soil.

Ants are social creatures, living together in colonies. And as in any well-ordered society, many ants have jobs, the details of which come to them naturally. These include entrance-blockers, seed-crushers, and honeypot workers (who store reserves of food for lean times).

Each colony has a queen whose sole responsibility is to reproduce. An ant queen can live up to twenty years and produce a community of some 500,000 ants.

Ants are one of my favourite insects. I admire the little beings for their wonderfully cooperative behaviour. When they invade my picnic basket, I try my best to leave them be. The ants feeding off my lunch will soon return to their colony with food in their bellies; they'll share this bounty by regurgitating it for the others. And, in a display of mutually beneficial behaviour, a hungry ant will beg for food by stroking the cheek of the food-gathering worker with his or her antennae.

Ants find their food of choice in our gardens: aphid honeydew. To get at the precious substance, an ant milks the aphid by stroking it with its antennae. Stroking relaxes the aphid and induces a flow of honeydew. Many gardeners are dismayed to learn that ants relish only the honeydew

In some countries, ants are used for suturing: The ant's jaws snap shut on the wound, and the body is snapped off.

In Europe, where prejudice against eating insects runs strong, the Swedes used to distill ants with rye, and the distillate was added to brandy to flavour it.

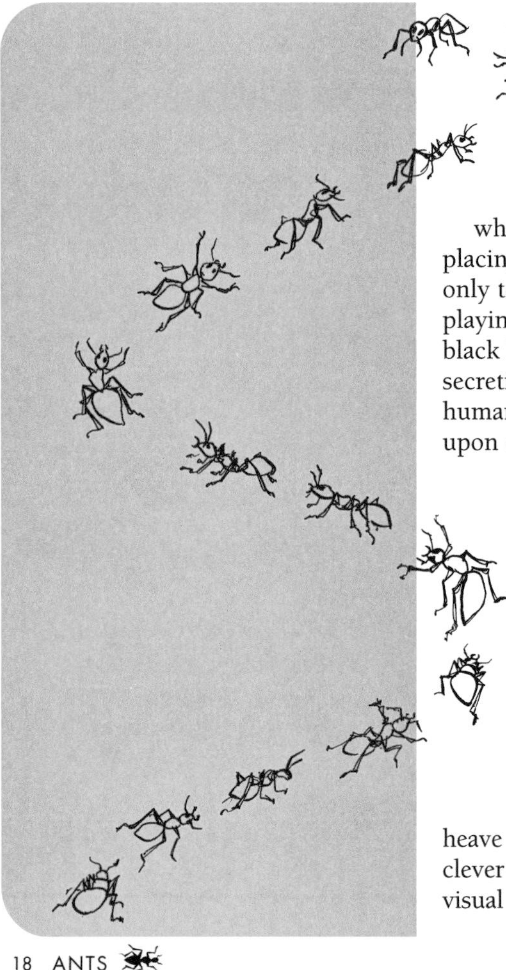

and not the aphid. Personally, I think it's a wonderful display of the dependence between species.

My new affinity for ants is quite a contrast to my childhood, when I loved to torment ants by placing obstacles along their paths. It only took one wood ant to stop me from playing this game. Wood ants (red back, black head and belly) spray a defensive secretion of formic acid that's visible to the human eye and produces intense stinging upon contact.

And while not all ants secrete formic acid, all use chemicals to mark their food and nest trails; an ability that helps other ants find their way. When I go hiking in the woods, trails are roughly marked and occasionally I veer off course. But when I catch sight of that precious piece of pink or yellow flagging tape, I heave a sigh of relief and think of those clever ants marking their trails with no visual clues at all.

Bees are among the most desired insects on a gardener's guest list. Without these industrious creatures to pollinate our plants, there would be no flower blossoms and our fruits and veggies wouldn't grow. Each of the 20,000 species of bees has four delicate wings, six legs, and an hourglass figure. They prefer flying to walking because, after all, they're busy bees with a lot of work to do.

Bees are responsible for pollinating over 150 crop species around the world. They're absolutely indispensable to the plant world, and thus to us. Nature has provided bees with a special tool to help them carry out their important duties. When their stomachs are full, bees keep on collecting pollen and nectar in special baskets located on their rear legs. I have a notion of bees buzzing, "That's one for the basket, and one for the belly." Some bees are picky, collecting only the pollen of a particular plant. And although these furry garden guests are vegetarians, don't think for a minute that they're pacifists. The girls – workers and queens – can give you a mean sting.

Bees, busy as they are, take time to dance. Returning to the hive after a tasty meal,

they use intricate dance steps to indicate the precise location of the food source to their friends, performing a series of circular moves and bottom waggling. The speed of the dance relates to distance – for example, a long flight equals many bum waggles. Bees also have a phenomenal sense of time; they know the opening and closing hours of virtually all the flowers that they feed on.

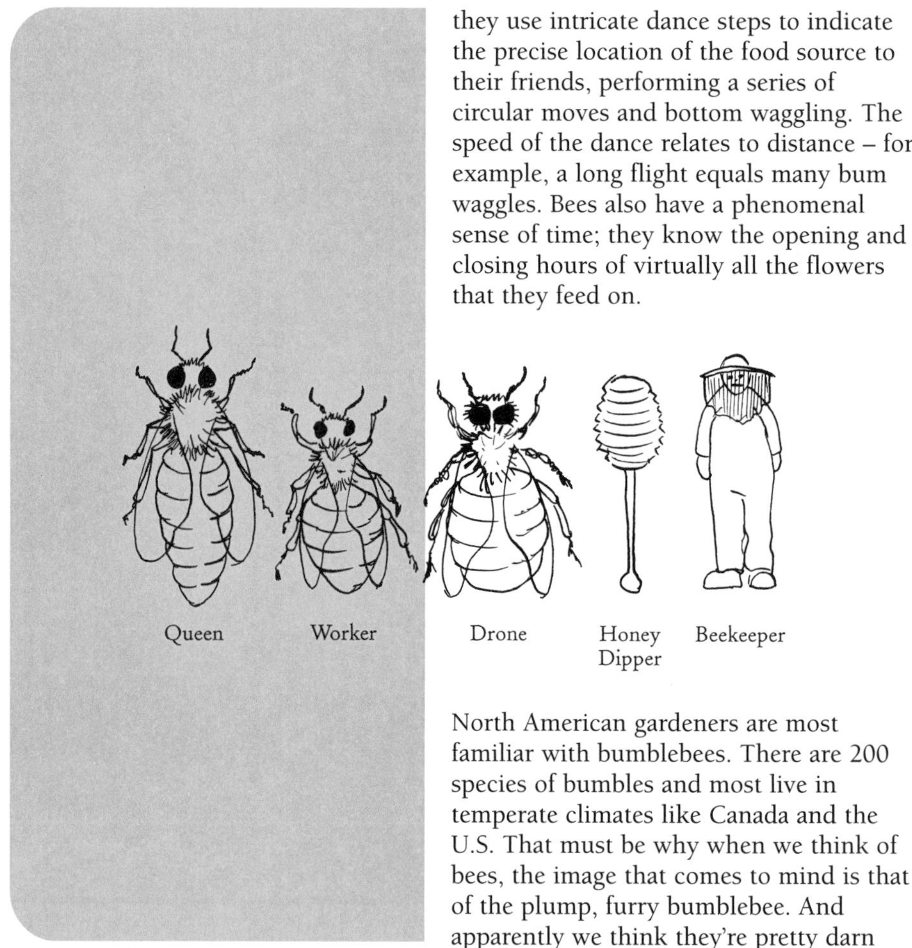

North American gardeners are most familiar with bumblebees. There are 200 species of bumbles and most live in temperate climates like Canada and the U.S. That must be why when we think of bees, the image that comes to mind is that of the plump, furry bumblebee. And apparently we think they're pretty darn cute too – how many cheerful kids in bumblebee costumes did you see at your door last Halloween?

There's a misconception that all bees reside in above-ground hives. Bumblebees often live underground; they make use of small animal holes to build their nests, and your garden supplies bumblebees with moss and leaves for insulation.

Probably because of their roundness, bumblebees don't execute their dance steps terribly well. But on the other hand, bumbles have extremely long tongues so they can suck up the nectar honeybees miss from the deepest blossoms.

Honeybees also enjoy most favoured bee status, except for the African honeybees that movie-makers love to sensationalize (*The Swarm*, anyone?). Fortunately these "killer" honeybees are still quite rare. Most honeybees enjoy a respected reputation for industriousness, and humans also appreciate their efforts to provide us with honey for our toast.

Throughout Europe it was generally believed that bees wouldn't thrive if important news were withheld from them – especially the news of when their keeper died. Bees were seen as the souls of the dead either returning to Earth or en route to the next world. So "telling" bees of important events like births, deaths, and marriages was a way of conveying news to souls no longer in a human body.

Sometimes relatives of the newly deceased would place black crepe on the hive for the period of mourning. Alternatively, hives were turned to face the other way as the coffin left the house.

Beehives were once thrown down from walls of besieged towns and forts to rout the enemy. This technique was used as recently as World War I when German soldiers tossed beehives into British ranks in East Africa.

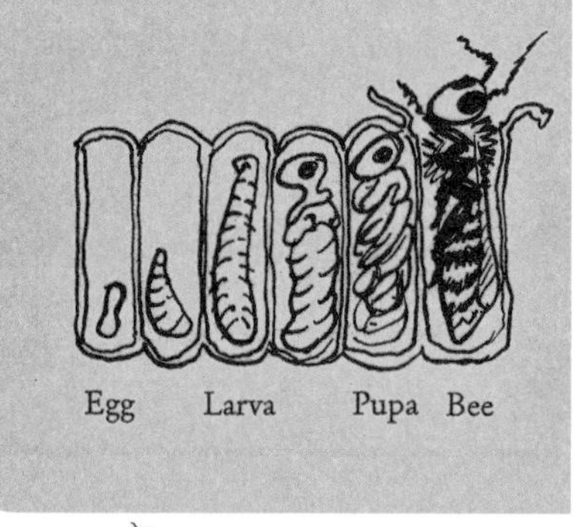

Like bumblebees, honeybees use their tongues to collect pollen and nectar. They store the precious substances in their little baskets and return to the hive to begin honey production. The honeybees take the nectar and, by adding secretions from their bodies, modify it into a yummy and energy-rich plant sugar: honey.

Honeybees live in nests of double-sided combs made of wax. The intricate cells of the honeycomb are octagonal, shaped exactly like stop signs. Given our fear of being stung, is this a coincidence?

Labour in a honeybee society is female driven. The queen is busy indeed. She lays up to 3,000 eggs in just one day. The workers (females) do the nest building, food gathering, caregiving, and just about everything else. The drones (males) do virtually nothing. They laze about demanding to be fed and waiting to mate. (That's bait I'm going to refuse to take!)

We've given wasps a bad rap. We often run in terror from them, but the majority of wasps don't sting: there are several thousand species of peaceful, solitary wasps, and just seventeen species of stinging social ones. Allowing our trepidation of the minority to taint our attitudes about all wasps is a shame, as they do much good in the garden.

These guests are accomplished hunters who particularly enjoy a meal of spider, caterpillar, weevil, cricket, or grasshopper. You may have noticed wasps for sale at your local garden centre. That's because many gardeners and farmers greatly value wasps as agents of insect warfare. Fruit tree growers frequently lose entire crops to weevils – they're usually first in line at the garden centre to buy large quantities of social wasps for releasing into their orchards.

Nearly every gardener has dealt with a wasp nest. These papery structures are the nests of social wasps like hornets and yellow-jackets. Social wasps can sting gardeners so it's wise to be very careful about approaching their nests. Of course, there are exceptions to every rule – although they're social wasps, hornets do not sting.

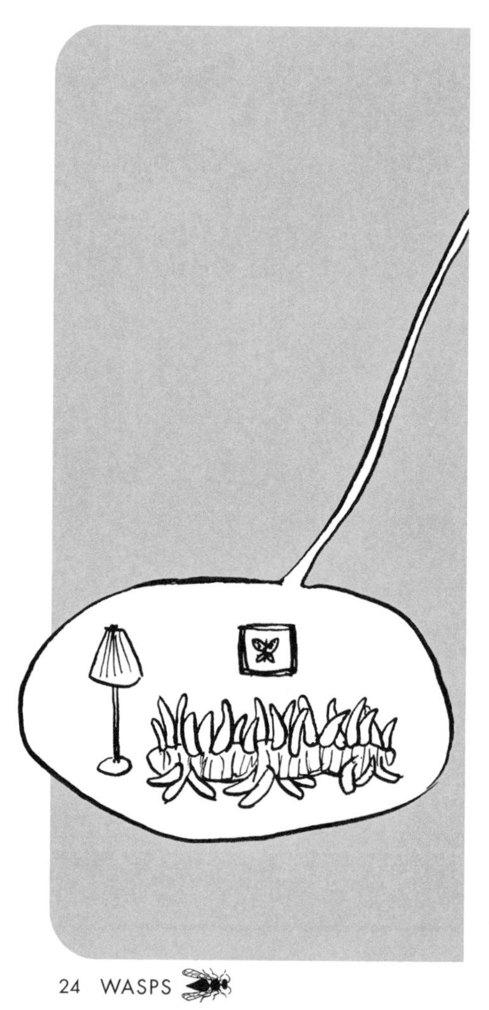

Hornet nests have a hard exterior with as many as 4,000 wasps living inside. They're extremely common in Canadian and American gardens. It would be hard to imagine the gardener who hasn't discovered a basketball-sized hornet nest hanging from a tree or under the eaves. When I was a little girl, my mother was afraid of hornets. Today I still get a chill when I so much as hear the word. But now that I understand that they mean me no harm, I'm slowly learning to relax when hornets buzz around the patio table; however, it's a gradual process.

It's the stinging tree wasps and other social wasps commonly lumped together under the term "yellowjacket" that it's perfectly logical to be wary of. These wasps look much like hornets, and they build their nests in similar locations. But tree wasp and yellowjacket nests are larger and more oval in shape than hornet nests. Look for yellowjackets in the fruit and berry areas of your garden – they love blackberries.

Solitary wasps live alone in burrows. They rarely sting humans unless provoked. If you have sandy soil in your garden, you've probably got a couple of resident sand wasps. They have a unique sense of decor – other insects are used as furniture. When a female sand wasp is ready to lay her eggs, she digs a hole in the sand and then goes in search of a juicy caterpillar or weevil to furnish the hole with. Once she find one, she administers a paralyzing chemical with her stinger and then, in an amazing show of strength, drags the unfortunate creature back to her burrow. As it lies alive but helpless, she deposits an egg on it and fills in the hole with sand, and then – voilà – her larva has a fresh food supply without ever having to leave home. And thus, the caterpillar sacrifices its life to help a new creature come safely into the insect world.

Ozark girls once carried little wasp nests pinned to the undergarments in the belief that this concealed nest would make them attractive to men.

You can recognize a wasp by its yellow and black body. But unlike the similarly coloured bumblebee, wasps are slender-bodied, and have even more clearly defined waists than bees. Take a look at an image of a tightly corseted Victorian woman and you'll see the resemblance between her unusually tiny waist and that of a wasp – hence the term wasp-waisted.

There are 3,500 species of cockroaches but we rarely see them. They're lightning quick on their feet and are nocturnal, two good reasons why cockroaches don't immediately come to mind as guests. There's a cockroach for every climate and therefore every garden. Though these little survivors thrive in warm, humid places, they also live in a huge variety of environments, including the North Pole.

Cockroaches grow as long as two inches. Hard forewings and a protective head shield make squishing a cockroach pretty difficult, and its flat, oval body allows it to scoot into tiny hiding spaces during the day. It hides from humans, but mostly from the birds, lizards, frogs, and other insects in your garden, all of whom relish a delightful entrée of cockroach. But well before a predator like you approaches, the cockroach's long, sensitive antennae warn it of your presence, giving it plenty of time to scoot for cover.

If you're like me, you grew up thinking that cockroaches are horrible, dirty things. There's some truth to this. Cockroaches can spread disease by coming in contact with our food, but in my opinion, they can be exceedingly beautiful while doing so. Most North American cockroaches are dark in colour, but my favourites are the tropical species. South Asian cockroaches are spectacular sights with their red, black, and yellow markings.

Cockroaches are garden recyclers extraordinaire. They're opportunistic feeders – they'll eat anything – and have specially adapted digestive systems to digest the cellulose in wood. Only cockroaches, termites, and some beetles can do this. Nature gave the cockroach a little pouch in its belly where a chain-gang of protozoans break down the cellulose to be used as nutrition. Some are born with these protozoans already in their gut. Others must acquire them by eating their parents' poop.

Not only are cockroaches fine racers, they can also fly. And as an added bonus, cockroaches are equipped with tiny adhesive pads on their rear feet that allow them to run along overhanging surfaces. Inspiration for the latest basketball shoes?

Still not convinced of the cockroach's fascinating habits? Like it or not, cockroaches are part of our lives. A cockroach female lays over 1,000 eggs before she dies (of natural causes!).

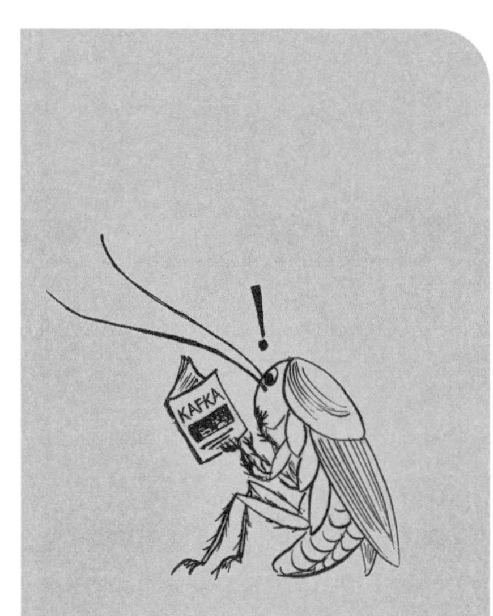

In Franz Kafka's classic novel *Metamorphosis*, the lead character is horrified when he wakes up as a cockroach.

Termites probably aren't on your guest list either, but they do sometimes hang out in the garden. Society knows the 2,300 species of pale, slender-bodied termites as destroyers of wood and builders of bizarre nests. Other societies think of them as a delicious source of protein.

Termites commonly build their nests underground but some construct landmark sized above-ground nests. The most incredible are seen in Africa and Australia, which can tower up to twenty-five feet in height and have been occupied by various termite colonies for decades. I guess termites know a good home when they see one.

In North America, nests usually are constructed inside dead trees, but occasionally you'll find a nest under or above-ground. Your garden provides termites with their necessary building blocks. They munch on wood and plant material, then hurry back to the nest site to add their feces to piles of other soil material. Termites also have organisms in their stomachs to help them digest wood cellulose.

Unlike me, termites need little personal

space. Up to a million live together in a single colony. It takes just a single reproductive female and a single reproductive male to found a colony. Once they mate and find the place they want to call home, the couple's wings drop off and they have no choice but to stay put. Termite parents make up for their lack of freedom by breeding their offspring as servants: workers, soldiers, and reproductives.

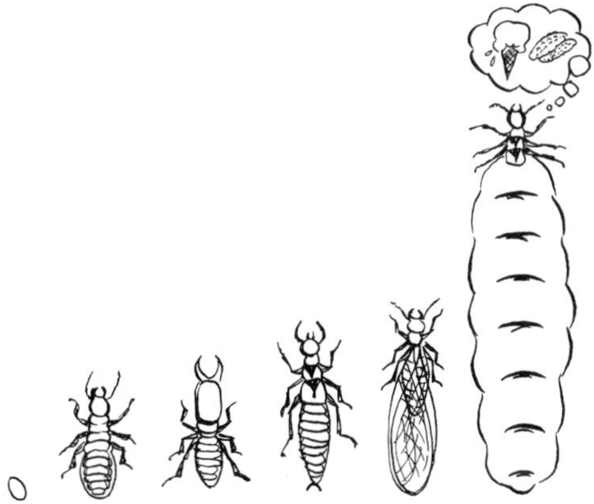

Ants and termites are mortal enemies. Ants march in formation to raid and snack on the inhabitants of a termite colony. If they're lucky, that is. If unlucky, the marauding ants are met by kamakaze termites who explode the contents of their stomachs all over the raiding party, sending them slip-sliding away, and no doubt utterly humiliated.

Egg Worker Soldier King Fertile Adult Queen

There has been considerable scientific study carried out on termite flatulence. In 1982, four scientists produced the first estimate of the annual production of methane by termites. One study, which was later contested, suggested that termite flatulence might be responsible for as much thirty percent of the Earth's atmospheric methane levels.

As a firstborn child, I feel sympathy for termite workers. They must have full-blown responsibility complexes. Not only do they play nanny to their soldier and reproductive siblings, they also provide food for the colony, and build and repair the nest. All that work and, at less than an inch in length, they're the smallest members of the family.

Queen termites are the biggest. At the height of her reproductive powers, a queen lays tens of thousands of eggs each day. I've never seen a real queen, human or termite, but photos of the pulsing, distended queen termite make me queasy. She's like a sumo wrestler at a buffet: ready to burst.

Beetles make up nearly half of all insects. Their astounding variety is no surprise given that over 300,000 beetle species populate the Earth. North American gardens are home to some 35,000 different species of beetle and their larvae, which are called grubs.

Although they vary in colour, size, and habits, all beetles have a distinctly separate head, thorax, and abdomen. They're frequent fliers, with hard forewings that meet over the midline of the abdomen. The delicate hindwings fold neatly underneath the forewings when not in use. A pair of antennae help the beetle tune into its environment.

Beetles can grow up to ten inches in length. Their six legs are used for swim-ming, jumping, digging, or running. Beetles have an intimate relationship with the plants in your garden; it's a relationship for which you may not be grateful. Beetles help gardeners by breaking down organic material to enrich the soil. The problem is their favourite organic material is often yours: leaves, trees, and fruit and vegetable crops. Some, like ladybird beetles, help out in a more desirable way by munching on other insects like aphids.

When I began researching this book, I noticed that most field guides don't mince words when talking about beetles. Words like massive, horny, exploitive, crushing, and predatory appear frequently. One species, the violet ground beetle, was described as a "frenzied murderer." All this negativity surprised me because, with the exception of weevils, I never heard a bad word about beetles when I owned a garden centre. It doesn't seem fair! I hope gardeners who read field guides won't come away with the notion that beetles are evil.

It's true that some beetles are legendary for their assertive behaviour, but they're also fascinating. Take the bombardier beetle. Bombardiers spray would-be predators with scalding hot chemicals. The beetle uses its internal combustion engine to spray the spicy mixture out of a little nozzle in its bum. The feisty creature hits its mark with startling accuracy, thanks to the ability of that bum to rotate up to 360 degrees.

There are so many remarkable beetles that it's nearly impossible to choose a few to highlight. Because this is my book I've included the beetles I find most notable:

weevils, ladybirds, fireflies, and June beetles.

Weevils

Weevils are small but they make up the largest group of beetles – there are over 60,000 species. I know from past experience that gardeners love to torment garden store owners with questions about terminating weevils. Weevils are probably the most common insect found munching on the fruits and veggies you've so diligently tended; while I hate to admit it, weevils can cause dramatic damage, particularly to strawberry and potato crops. They are easy to distinguish from other beetles because of their long heads, bent antennae, and tremendous beaked snouts.

Ladybirds (a.k.a. ladybugs)

Tiny, round ladybird beetles are widespread throughout North America. Most of us think of ladybugs as small red insects with black dots. This is mostly true – they are small (a quarter of an inch), but many ladybirds have yellow wings instead of red. No matter what the colour, they have one large black spot over the join between their wings, and three or more smaller spots on each wing.

Ladybird beetles were considered to be an excellent remedy for colic and measles. They were also used for toothaches: one or two were mashed and stuffed into the cavity of the tooth to ease the pain.

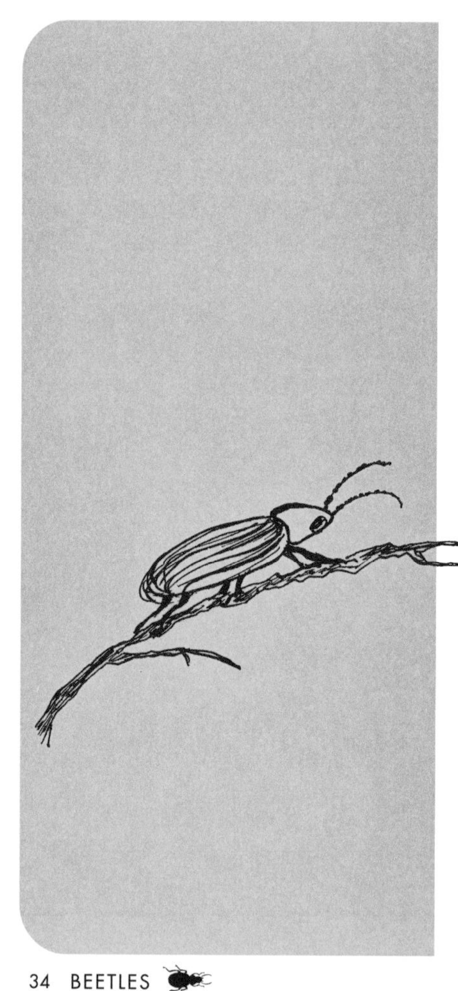

Have you ever seen bunches of tiny, brilliant yellow eggs in your garden? These are ladybug eggs, which hatch in just three or four days into teeny black crocodile-like creatures – the ladybug larvae.

Considerate gardeners don't use pesticides when they have aphid infestations. When they simply can't take it any more, these gardeners bring in the big guns: ladybugs. The little ladies are revered among some gardeners for their voracious, aphid-fuelled appetites.

Fireflies

I love the quaint name my French friends have for these glowing creatures: *feu follet*, which means spooky fire. I'm not sure why in English the suffix fly is used, since technically they are beetles. Male fireflies have the same wings as other adult beetles, while females never lose their larval form and are referred to as glowworms.

Fireflies and glowworms are nocturnal. If you're fortunate, you'll see them flashing at night among your perennials. But sadly, the sight of a firefly or glowworm's lantern on a sultry summer evening is becoming increasingly rare. I believe this is due to the heavy, and unnecessary, use of pesticides. Hopefully the surge of interest in organic gardening will help brighten these magical creatures' lamps once again.

Scarab beetles (a.k.a. June Beetles)
There are over 1,000 species of scarabs living in North America, mostly in the southeastern United States. With its large, brightly coloured head, and square, glossy iridescent body, a scarab is easily recognizable in your garden. Scarabs eat a varied diet – they're fruit eaters, pollen drinkers, and connoisseurs of dead things. Their plump white larvae are attracted by light. In the early summer you may find them attached to your windows.

In Egypt, about 600 B.C., a carved image of the scarab beetle was placed next to the heart of a mummified body. The flat side of the Scarab was always inscribed with a spell from the Book of the Dead, to prevent the heart from testifying against its owner during the post-mortem weighing of the heart before the divine judges.

THE BEAUTIFUL

Moths & Butterflies

I used to think the distinction between moths and butterflies was clear: moths were beige-coloured and boring, while butterflies were exciting works of art. Moths slept in the daytime, waiting dumbly until nightfall to beat themselves to death against hot light bulbs; butterflies spread their luminous wings by day and sensibly retired in the evening. Boy, was I wrong.

Not all moths are night fliers, and many are stunningly coloured. This is so often the case that we frequently mistake them for butterflies. In fact, you may be surprised to learn that the 13,000 species of North American moths and butterflies share more characteristics than not. These gentle vegetarians have a prominent, tube-shaped mouth for sipping flower nectar and for pollinating our plants. Like bees, many eat from only one food source, preferring one type of flower in your garden to another. Flower nectar has little protein, so, sadly, adult moths and butterflies live only a few weeks.

How the heck do we tell them apart? Admittedly it's difficult, but some general rules apply. Note that I say *general*. Moths have heavy, furry bodies, feathery antennae, and keep their wings spread while resting. Butterflies are smaller, have clubbed antennae, and bring their wings together at siesta time. Does this help?

Have you ever touched a butterfly or moth's four wings? If you have, you've seen the flour-like dust that rubs off onto your hands or clothing. This dust is made up of tiny scales that cover the creature's wings, head, and legs in a tight, overlapping pattern. Up to 10,000 scales per wing give the moth or butterfly its unique pigment and pattern. If you were to scrape off the scales – which of course you mustn't – you'd expose the transparent wings underneath.

The stunning colours and patterns are a signal to predators to stay away: the more beautiful the butterfly or moth, the worse it tastes. Predators like birds, spiders, and beetles quickly learn that these brilliant creatures make a most unpleasant meal.

Colour also helps to control their internal thermostats. Just as we avoid working in the garden under a hot summer sun while wearing dark clothing, so too do moths and butterflies. Dark colours absorb more heat from the sun. Typically, the hotter the

Jeff Bloomquist and Steve Bathiche have come up with a way to harness the neuromuscular reactions of insects to control model cars. They initially began with a cockroach in the driver's seat of the model car, but according to Bathiche, "cockroaches only fly short hops, and they don't steer that well." Now a hawkmoth sits in the driver's seat. Hawkmoths "are much more elegant fliers. When they want to turn, they lean in the opposite direction." Indy drivers, watch out!

climate, the less black you'll see on a butterfly or moth's wings.

Every gardener loves to see a colourful butterfly flitting among the annuals and perennials. But if you're like me, your preconceived notions about moths may have kept you from honouring their presence in your garden. Now that I know more about moths, I welcome these maligned creatures into the garden just as I do their butterfly cousins. I hope you will too.

The larvae of monarch butterflies are unable to survive frosty temperatures. Instead, the monarchs fly south to set up huge larval colonies in Mexico and the southern U.S. Considering the lack of protein in their diet (flower nectar isn't terribly nutritious), it's nothing short of miraculous that they travel the long distances they do. Monarchs can travel over a thousand miles in less than ten days.

I f you're an observant person, you'll see minute white spots on the trees in early summer. These are moth and butterfly eggs. Two or three months later, the eggs hatch into the larvae we know as caterpillars.

Tiny critters, often just a few sixteenths of an inch long, caterpillars vary in colour – red, black, yellow, white, or a combination – and most are hairy. Once hatched, they head out immediately in search of greenery to snack on. And this usually means they will set a course leading straight for your garden.

Over the next two months, caterpillars are active participants in their version of a biathlon: eating and sleeping. During the day they sleep in temporary nests spun from their own silk, and at night they emerge to feast on your most tender plants.

It's a frustrating experience to have tender seedlings mowed down by caterpillars, but you can console yourself with the knowledge that you're helping these future moths and butterflies to survive. Eventually they'll reward you by brightening your world with their beauty.

CATERPILLARS

caterpillars

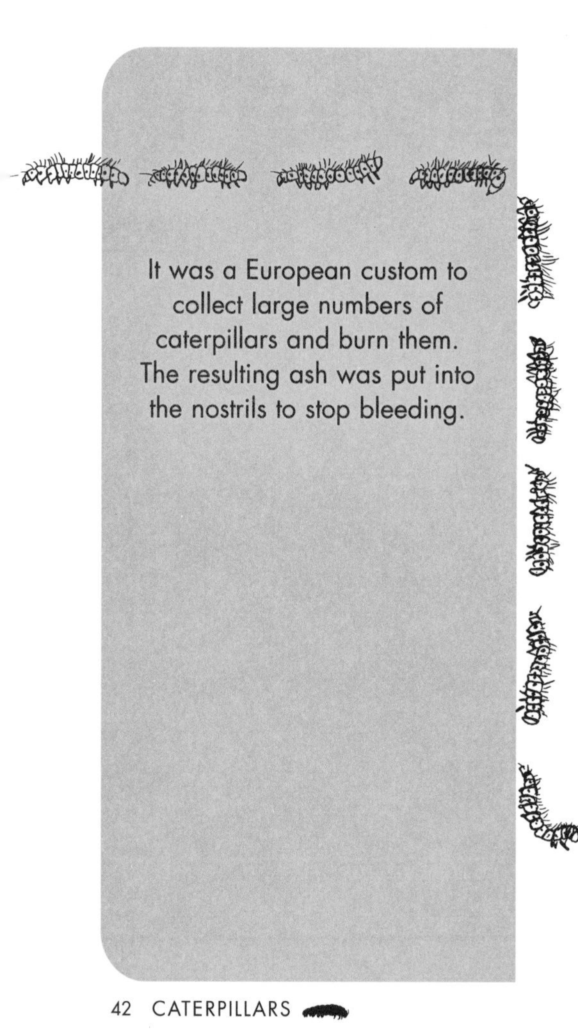

It was a European custom to collect large numbers of caterpillars and burn them. The resulting ash was put into the nostrils to stop bleeding.

When the weather gets nippy in late fall, moth caterpillars will build a cocoon high in a tree. The cocoon is made of silk supplemented with leaves. Butterfly caterpillars develop into a chrysalis instead. The chrysalis (or pupa) is an additional stage between larva and adult that moth caterpillars don't go through. The chrysalis burrows into the ground or attaches to a tree for the winter.

Come spring, moth caterpillars inch down from their cocoon to the ground, often in long line-ups, and bury themselves in the earth. Butterfly chrysalises can hang out where they are. In time, they will emerge as moths and butterflies, the sight of which is guaranteed to gladden the heart of the most jaded gardener.

No book on garden guests would be complete without reference to leaf miners, a type of moth larvae. Gardeners talk about these little insects as though they are serial killers. Too bad – the leaf miner deserves our admiration for its ability to construct intricate tunnels between the outer layers of the leaves in your garden. Not only that, leaf miners are excellent housekeepers. As the industrious insect tunnels across the leaf, it eats the resulting plant debris, leaving no mess, unlike that left by human miners. Leaf miners are also picky eaters. There is a type specially adapted for most of the plants in your garden.

We are blessed with over 5,000 species of damsel and dragonflies worldwide. Never heard of damselflies, you say? I hadn't either until researching this book. That's because damsels are often lumped in with their larger, more robust relations, the dragon-flies. It's easy to understand why: both species have iridescent, richly veined wings and brightly coloured legs and bellies. Dragons and damsels have inspired count-less works of art through the ages and have made generations of greenthumbs happy because they eat aphids with gusto, along with beetle larvae (grubs).

If you look closely, you'll notice that damsels are petite versions of dragons. Despite their dainty stature, it's likely to be a curious damselfly that comes in close to check you out while you're weeding. For all their robustness, dragons are easily startled and shy around humans.

I was quite sad to learn that neither damsel-flies nor dragonflies have the ability to tuck their beautiful yet fragile wings under hard-ened forewings like other insects. Exposed wings are a serious problem, for it means these lovely creatures cannot hide in tight spaces without tearing their delicate wings. The poor dragonfly must hold his wings horizontally across his body at all times, while the damsel can at least hinge hers vertically when she needs to.

I'm one of those people whose core body temperature seems to be perpetually below normal, so the damsel and dragonfly's hypersensitivity to cold earns much empathy from me. Because body warmth is so critical to the survival of damsels and dragons – they cannot fly if chilled – the little creatures use the sun to warm up. No doubt you've been treated to the splendid sight of an iridescent dragon or damsel basking in the sun in your garden. These beautiful creatures aren't just lazing around catching a few rays; they're warming their flight muscles.

With luck, you'll also have the chance to observe dragons' and damsels' ritualized mating habits. Airborne males of both species perform complex aerobatic moves – spirals, zigzags, dives, and circles. It's a lovely thing to witness.

As with humans, the competitive instinct varies among dragon males. Some go so far as to knock aside another male already having his way with a female. The newcomer scoops out his predecessor's sperm

The male dragonfly has an amazing penis. Not only is it inflatable, it's articulated too.

There is an American superstition that damsel and dragonflies will sew one's ears together (and lips, nostrils, eyelids, even fingers and toes). Thus they are sometimes referred to as the devil's darning needle.

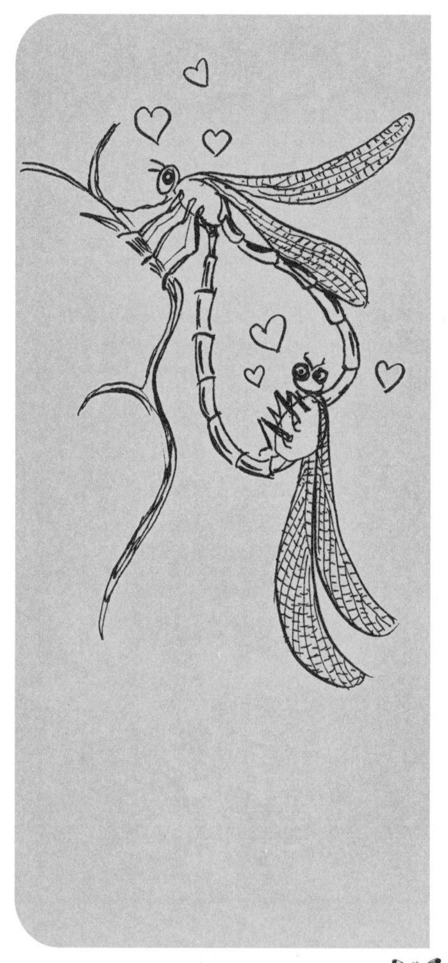

before mating with the confused but perhaps flattered female. Both damsels and dragons mate in mid-flight by forming a tandem – no small feat when you consider that they can fly sixty miles an hour. The male holds the female with his legs and locks special hooks on his belly onto her head (dragonflies) or thorax (damselflies). This flying sex act can last for several minutes to several hours. (Gosh, and I can barely walk and talk at the same time.)

I believe that gardeners are a kind lot. Our efforts to green the world can make up for some of the terrible impact we humans have on our environment. Acid rain and other pollution have done much damage to the traditional habitat of many life forms. And human encroachment by filling in marshes and ditches is shrinking what pristine environments remain. I hope you'll agree that we have a responsibility to delicate creatures like dragon and damselflies that need our help to survive.

Give these inspiring objects of beauty a peaceful home by inviting them into your garden. Build a pond in your green space and leave soggy areas alone whenever possible.

Living in British Columbia, I admit to never having encountered a mayfly. They congregate in eastern parts of North America, where there are nearly 600 species. If I ever garden in eastern Canada or the U.S., I'll keep my eyes peeled for these unique insects. I'm told it's easy to recognize an adult mayfly, which has a large pair of clear, shiny wings and two smaller hind wings. A mayfly's body is less than an inch long, pale in colour, and has two extremely long tail filaments.

Their life story is a lesson in patience and impermanence. They spend up to two years as larvae, waiting for their debut as reproductive adults. But once they do emerge, these beauties experience sexual maturity for just a few days. Not only that, during this short time their bellies are empty. You can tempt them with all the juicy insects you wish, but mayflies won't take the bait – they're reducing their body weight for more urgent business like searching for a mate.

I'm fascinated by the mating habits of all creatures. From elephants to earthworms, each has its own special way of getting it on. As for mayflies, huge swarms of males

Many mayfly species just spew their eggs into the water as they fly above the surface. Some do not actually lay their eggs; their egg-laden abdomens simply break off and fall into the water.

Mayflies are attracted to light. In cities and towns along the Great Lakes and the Mississippi, they can accumulate in huge drifts that must be cleared away with front-end loaders and dump trucks.

take to the sky in May and June to perform ritualized mating dances. They fly up and float down, fly up and float down, over and over again. When a female happens into the swarm, a male seizes her from below, locks his forelegs around her body, and inserts his admirable double-barrelled penis into her equally admirable double vagina.

Mayflies leave this world in one big orgiastic party (make sure to get yourself on *their* guest list). Larvae emergence, mating, egg laying, and the subsequent death of the adults all take place within a matter of days. You may have seen a mating swarm or the final piles of dead mayflies littering the surface of your pond if you have one.

A sad end? Perhaps. I prefer to think of it as an example of living in the moment – like there's no tomorrow.

THE MOTLEY CREW

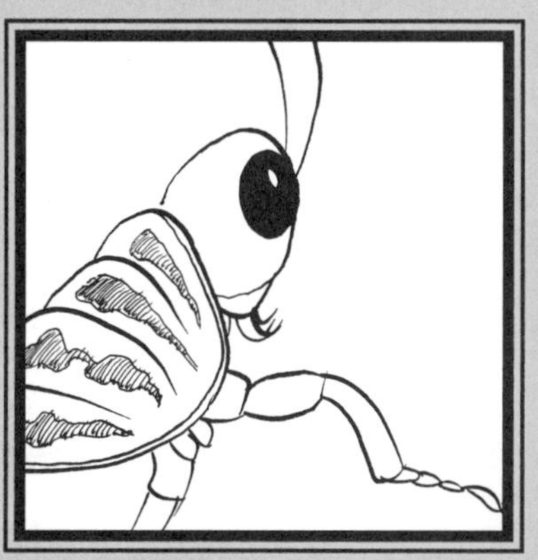

the motley crew

That's praying, not preying. The name comes from the appearance of the mantid's forelegs folded as if in prayer. There are nearly 2,000 species of mantids, but we North American garden-ers get to see less than a dozen. Too bad, because praying mantids are excellent crowd controllers. They eat guests you may find unwelcome, like caterpillars and aphids.

If you live in the eastern U.S. or central Canada, you'll find mantids on shrubs and trees in open areas like woodlands. They're about two inches long and have a tiny, triangular head atop an elongated neck, long skinny antennae, and two prominent eyes. The mantid is an accomplished hunter. Its forelegs are oversized, powerful, and ideal for grasping prey.

The mantid and the possessed child in the movie *The Exorcist* have a lot in common: both have a head that can swivel completely around. It can be quite disconcerting to see a mantid's body moving in one direction and its head looking back in the other.

The praying mantid is a devotee of the practice of patience. When hungry, it waits in prayerful contemplation for prey to happen by, at which time it unfolds its powerful, armoured forelegs and seizes the unsuspecting victim. The mantid usually isn't a picky eater; just about any insect will do.

Mantids can run, or even fly, from other insects, birds, or other predators. But the mantid's most remarkable defensive device is its ability to scare the heck out of would-be predators. The seemingly fearless mantid rears up on its hind legs, lifts the muscular, spiny forelegs above its head, and spreads its electrically hued wings. This gob-smacking vision is a shock that gardeners, as well as insects, and other animals are unlikely to forget.

My friends Henk and Molly once witnessed an interesting showdown. On a walkabout in France, they saw two small

You may have heard that female mantids sometimes eat their mates during copulation. Scientists say this rumour is greatly exaggerated – a female will eat a male only if she's starving. Small consolation to her companion.

Four hundred years ago, a man named Wang Lang in the province of Shantung, China developed a type of kung fu based on the postures and movements of the praying mantis.

Try giving the female praying mantid the run of your house. She'll eat out of your hand and kill insects for you. In just a few minutes, she'll devour a fly, except for the wings, which she'll drop on your floor. While not the tidiest of housekeepers, the female mantid is fastidious in her personal hygiene habits. She may spend up to fifteen minutes cleaning her face and legs after eating.

creatures moving violently on the ground near a hedge. One was a finch, and the other, well, they weren't quite sure, but it was standing up and flashing pink and blue neon wings. The bird eventually backed off and my friends were astonished to see the winner of the battle was a praying mantid. Although this event took place twenty years ago, they both remember it with great clarity.

Gardeners who live in mantid territory should take some time to go in search of its unusual egg clusters. These are found on tree branches or in the grass. Mama surrounds up to 200 eggs at a time with a frothy, sticky mucus that hardens to protect the eggs during the winter months.

Grasshoppers & Crickets

Although they resemble mantids, grasshoppers (which include crickets) are really quite different. Grasshoppers and crickets are distinguished from mantids by their much larger heads, as well as their hind legs, which are more powerful than a mantid's. The delightful clicking sounds made by these little musicians are among their other notable features. When I hear a cricket or grasshopper clicking away in my garden, I know summer has arrived.

There are over 1,000 species of crickets and grasshoppers living just about everywhere in North America except in very cold areas. They make their homes in garden vegetation, the ground, under rocks, and in decaying wood. In the winter, they take cover in burrows, waiting for the warmth of spring.

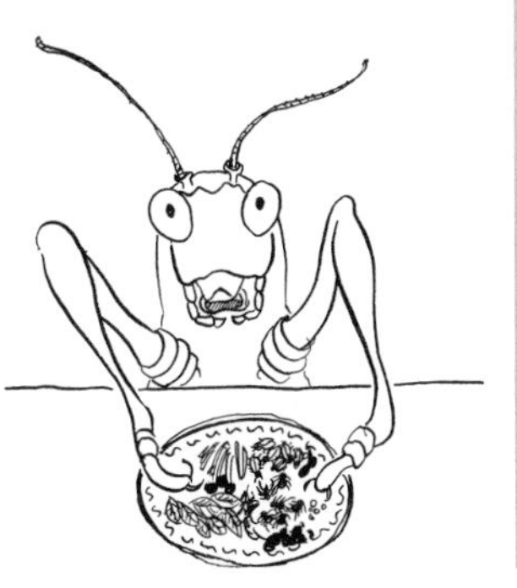

It's easy to find a menu that suits these garden guests –

Crickets make a great musical thermometer. Count the number of chirps a cricket makes in fourteen seconds. To measure the temperature in degrees Fahrenheit, add forty and you'll have the current temperature. To find the equivalent in degrees Celsius, add twenty-two.

they'll eat plant material and/or other insects. Like virtually all creatures in our gardens, grass-hoppers and crickets have a strategy to avoid becoming someone else's meal. Some choose to blend in with their surroundings, while others have foul-tasting bodies that they advertise with brilliant colours. Predators, recognizing the distinctive hues, avoid the distasteful critter.

Grasshoppers and crickets are athletic insects that can evade capture by catapulting into the air. Although they fly only a short distance, both species use the airtime to make their famous clicking sounds and to flash their resplendent wings. The predator is momentarily taken aback by the movement, the clicks, and the colours. By the time it recovers and tries to follow the trajectory, the cricket or grasshopper is silently resting nearby with wings neatly folded. Camouflaged once again and, no doubt, feeling quite smug at its trickery.

If caught, the grasshopper or cricket kicks at the predator with spiny hind legs. I've also been told that they sometimes spit up their stomach contents with the hope of gaining freedom by grossing the predator out.

Males are roving minstrels who court females with lovely mating music and elaborate dances. Mating songs can be quite loud, heard by us humans at some distance on still summer evenings. Female crickets and grasshoppers use special ears to listen to these courtship songs. These ears are found on the underbelly (on grasshoppers) or on the legs (of crickets) and are made of thin membrane with sound receptors attached. Hearing the males sing for the first time each year reminds me of the passage of time. Grasshoppers and crickets help me to be mindful of the rhythm of my life and how entwined it is with that of nature.

They also are part of a loving ritual. My young friend Amadea and I buy live crickets from the pet store, where they are cruelly sold as snake food. Together, we release the lovely jumpers in her garden. Amadea's squeals of delight, her natural fascination, and the gentleness of her tiny fingers give me a wonderful sense of hope. If we help children to respect and treasure the interconnectedness of all life forms, the world will be in good hands.

Just before the onset of World War I, a German scientist showed that female crickets were attracted to a telephone if a male of their species was singing at the other end of the line. This proved that females respond to the calling song alone without assistance from chemical or visual signals.

Not only do they look like twigs, branches, and vines, the 2,500 species of walking stick insects also behave like vegetation. Do you remember playing "Let's Pretend" games as a child? Pretend you're a tree. Pretend you're a leaf. Well, that's the curious life of a walking stick. Waving to and fro, they try their darnedest to look like a twig swaying in the breeze.

I've certainly looked, but I've never seen a walking stick insect in my garden. Most North American gardeners haven't either. Or have we? Perhaps they're just too well camouflaged. Given their twig-like appearance, and the fact that walking sticks seek out gardens with lots of shrubs and trees, it's no wonder that it's so difficult to discover one.

Like many other insects, walking sticks spray stinky stuff from their bowels at predators. And who can blame them? They don't have wings, and they live in the open, so they need *something* to protect themselves.

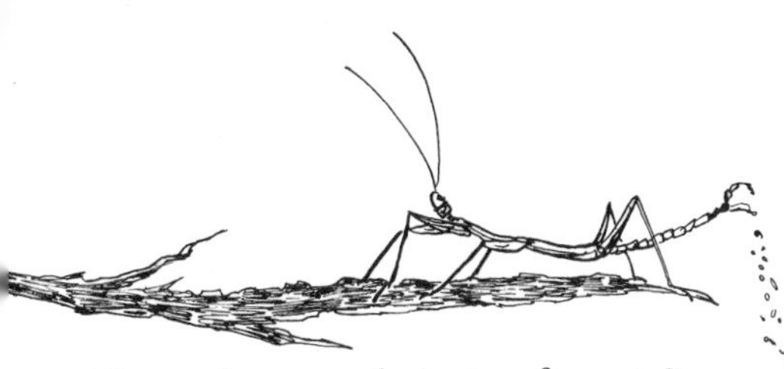

Remember my admiration for ants? Here's yet another reason I love them. When female walking sticks release their eggs, they just let 'em fly from high up in the trees and down to the ground. Ants rescue the eggs by carrying them back to their underground nests. But the ants don't eat them – they feast only on a large edible appendage attached to the hard-shelled egg. The ants also don't interfere with the tiny, newly hatched walking sticks as they make their way from the ant nest up into the trees, where they grow up in relative safety.

A walking stick may keep his mate from being inseminated by other males by coupling with her until long after he has passed his sperm into her genital tract, like a living chastity belt. The known record for prolonged copulation by insects (79 days) is held by a pair of walking sticks.

Thrips is a peculiar name that ends with an "s" in both the singular and the plural. Onions and garlic are the preferred hosts of the 5,000 named species of thrips (scientists say there are thousands more yet unnamed). This isn't great news for many of you who've had other crops like cabbage, celery, tomatoes, beans, cucumber, and pineapple destroyed by thrips.

You'll find these voracious insects on your plants, and under bits of bark, or decaying wood. You'll also see them on your flowers, using their tiny sucking tubes to feast. But, like all of us, they have redeeming qualities: they feed on scale insects and mites.

You've probably noticed evidence of thrips making a visit to your garden. When a leaf that's been snacked on by thrips grows, the holes also enlarge, leaving empty spaces on the surface. This damage appears as silvery patches or streaks that shine in the sun. It's actually rather pretty, I think.

Thrips are among the most petite guests in our gardens. Some are less than a sixteenth of an inch in length. They arrive

in groups, wearing yellow, brown, or black. Under a microscope you can see that they have two pairs of long, delicately fringed wings resembling ostrich feathers.

For some odd reason, thrips are especially active during thunderstorms. If you have exceptional eyesight, you may get the chance to watch thrips pretending they're hang-gliders. They like to crawl to the top of a plant and then leap off, spreading their wings like a sail to float gently downwards.

Thrips have unusual feet. They're hoof-like in shape and have a membranous bladder that the thrips inflates or deflates at will. These sticky balloons help thrips negotiate slippery surfaces, such as the petals of your flower blossoms.

EARWIGS

earwigs

One of your garden guests will be the earwig. The 1,200 species are energetic little things, dark brown in colour and, like so many insects, they have rather flattened bodies. Their rear wings are large and delicately folded in accordion pleats, and the short but hardy forewings lay over top to protect the intricate folds.

The thing you'll notice straight away about an earwig is the small but fierce-looking set of pincers protruding from its rear. The earwig uses this device to unfold its wings, act as antennae, seize prey, assist in the mating process, and defend against predators. Whew! The pincers of males are curved, while those of females are straight and close together with overlapping tips.

Although common visitors to our gardens, we tend to see earwigs only when disturbing rocks, leaves, and wood piles. They like dark places and prefer night action.

Earwigs aren't hard guests to please. Their favourite meals include other insects, dead and living plant material, even flowers, fungi, berries, and rotten fruit. They help to return nutrients to your garden soil, so don't get too bent out of shape when you find an earwig nibbling on your raspberries.

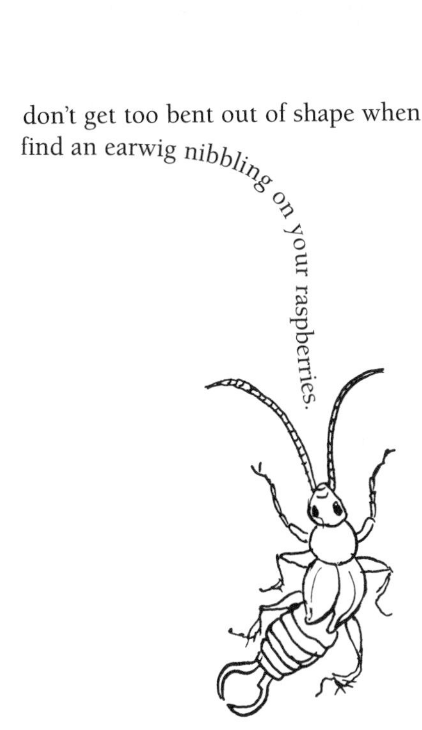

Earwig is from the Old English *earwicga: ear* because the insect was thought to crawl into the head via the ear, and *wicga* (or wig) for the wiggling movement. Some people also thought that spitting into the ear could encourage an earwig to come out.

THE IRRITATORS
the.
irritators

Fleas live everywhere on the planet, but they thrive where summers are cool and humid, and winters are mild and wet. There are 1,800 species of the little critters but because they're less than an eighth of an inch long, we tend not to notice them. Nevertheless, I can virtually guarantee there are fleas in your garden.

Fleas don't contribute to the world in the same ways as other insects – they don't recycle decaying material, nor do they control aphid populations or pollinate flowers. These wingless jumpers eat only blood and hang out in your green space waiting for a host body. This could be a rodent like a field mouse, a bird (fleas love swallows), or your cat or dog. Even gardeners make a tasty snack.

If you can get one under a magnifying glass, you'll see that a flea is covered with backward-facing bristles. These help the minuscule being anchor itself on the host's body. Given their size, it's hard to believe fleas can jump nearly twelve inches, but thanks to their powerful hindlegs they do so with ease.

My mother, and the other exemplary

housekeepers of her generation, will be horrified to read that fleas are pretty much permanent guests in homes with companion animals.

My two cats, Tsewang and Rinchen (the boys), carry around a flea or five all summer long. Tsewang in particular is a flea magnet, and unfortunately he suffers an allergic reaction to his little tormentors. Unless I diligently comb and pick the fleas off his furry body, he is soon covered in ulcerated sores. To help out, I practise a capture-and-release strategy with his fleas: once caught, they're carried between my fingers and set free in my neighbour Craig's garden.

You may think this a ludicrous idea. But the fleas are perfectly content to hang out in the shady, sandy soil in my garden, where they can live for months without feeding. So, yes, I release them and they bide their time until poor Tsewang happens by again.

Kitty won't agree but I think it's fascinating the way fleas pierce skin to access blood. The flea's tiny head contains a hammering device attached to a piercing stylet. When the flea finds a suitable place

In medieval Europe the Black Death, or plague, killed millions of people. Named for the blackened blotches on the afflicted person's skin, the Black Death was transmitted to humans by rat fleas. The flea would bite a rat that carried the plague bacteria, and then when the flea bit a human (as they often did in those unsanitary days), it spread the bacteria into the bite wound. The person was doomed.

to feed, it cocks the hammer, and positions itself head down and rear up. The hammer is sprung and slams against the victim with enough force to drive the stylet into the skin. Unfortunately for the host, this happens time and time again, rat-a-tat-tat, until the flea locates a good supply of scrumptious blood.

Flea circuses were once extremely popular in Europe. Circus fleas were sold, already harnessed, with steel or silver collars. Thomas Moffatt, a contemporary of Shakespeare's and author of the first English book on entomology, was very impressed with an English flea circus, where the flea dragged a gold chain and lock behind it.

Ralph Waldo Emerson wrote, “A fly is as untameable as a hyena.” When flies congregate on the inside of my screened windows, Emerson’s words often come to mind and I have to smile. Actually, I don’t really mind flies as long as they aren’t eating or pooping on my food.

I try not to get too worried about the little creatures, for in North America there are more than 15,000 species. Flies have two membranous wings and a pair of oversized eyes. Once upon a time flies had four wings, but since then two have evolved into useful balance devices called halteres.

Flies are universally considered to be carriers of germs. That’s because they’re scavengers who eat nasty, decomposing things. Flies spread malaria (specifically mosquitoes), contaminate food, and in the tropics affect human health in occasionally gruesome ways. On the other hand, flies get rid of much of the world’s decaying matter; think of them as recycling experts. We should also give flies credit for the work they do in our gardens to pollinate flowers and crops – second only to bees, wasps, and ants.

Many South American Indians say that spirits and demons sometimes assume the shape of insects. The Auraucanian Indians believe that departed tribesmen, chiefs in particular, take on the form of horseflies. When horseflies enter a village where someone is sick the Indians begin to wail as if death had already occurred, thinking that the souls of the person's dead relatives have come to take him or her away.

Talented pilots, flies can cruise (and land) upside down and backwards. Those halteres tell the fly if it has strayed off course and how fast it's cruising. Flies can hover too, and they have exceptionally high wing beat frequencies – hundreds of beats per second.

Although only some adult flies are predatory, virtually all fly larvae (maggots) are carnivores. Some gardeners will be pleased to learn that maggots eat aphids, beetles, and even slugs and snails.

Some of the fly species North American gardeners are most familiar with include mosquitoes, craneflies, fruit flies, blackflies, horseflies, robber flies, bluebottles, houseflies, deerflies, and blowflies.

MOSQUITOES

mosquitoes

Many evenings after being driven indoors by these whiny, biting visitors, I've longed for a world where mosquitoes are always guests in someone else's garden. Even the revered pacifist, His Holiness the Dalai Lama, admitted to once ending the life of a mosquito. As he swatted it, he prayed, "Go to the Pure Land!"

Scientists believe that as many as fifty percent of all human deaths can be attributed to disease carried by mosquitoes. Personally, I think we humans share some of the responsibility. We've sprayed huge swaths of the tropics with chemicals to eradicate malaria mosquitoes. And doctors urge travellers to gulp down anti-malaria tablets. What have these plots accomplished? In my view, we've only forced mosquitoes to become more resistant and to carry stronger forms of pestilence. I've travelled several times in malaria red-zones. In the early nineties off the coast of Cambodia, the threat of being bitten by a malaria-carrying mosquito was serious. I chose, however, to use a mosquito net at night, to wear repellent, and to burn

Capuchin monkeys in the forests of Venezuela protect themselves against mosquitoes just as we do – by rubbing repellent all over their bodies. They poke around in termite mounds and bark to find the four-inch-long *Orthoporus dorsovittatus,* a millipede that secretes the benzoquinone chemicals that repel jungle mosquitoes. The happy monkey massages the unhappy millipede into its fur and mosquitoes go elsewhere for a meal.

mosquito coils to fend off the little bloodsuckers. In the end, I had far fewer bites than my tablet-swallowing companions and my hair didn't fall out from the pills like theirs did!

Mosquitoes, who are members of the extended fly family, have a good side too – they play a useful ecological role. Their larvae are filter feeders that eat tiny algae. In turn, the larvae are eaten by fish and, after hatching, by birds.

Only female mosquitoes bite. The role of males is to drive us nuts with their whining – er, that would be love songs. A male in the mood for romance serenades females in the area by rubbing his furry antennae together, but unfortunately, our clumsy ears don't catch the subtleties of his tune. I suspect that some nights I entertain entire marching bands of males in my garden.

So what do males eat if they don't bite and suck blood like the females? They use their piercing stylet to suck fruit and plant juices instead. Females have a similar, but enhanced, six-part stylet that they prefer to use on beings like us. No wonder we feel the bite of these gals so keenly.

The cranefly resembles a vastly oversized mosquito – by comparison, the cranefly's skinny body is usually about an inch long. Craneflies fall into the fly family and they live all over Canada and the U.S., especially in damp areas. Some cranefly larvae live in the ground and nourish themselves on dead plant material, so they don't deserve the reputation gardeners have bestowed upon them: pests, not guests. But a variety of cranefly larvae called leatherjackets can certainly be a problem for those who suffer from best lawn on the block syndrome.

Leatherjackets also love to nosh on your fall cover crop. But if your cover crop is weeds, then just think of all the time you'll save!

The name leatherjacket suits these wriggly youngsters perfectly. The skin on the outside of the plump larva is toughened like a tanned hide, providing protection from pointy bird beaks until the larva reaches its winged adult stage in the summer.

Adult craneflies and I share a horrible secret. As a child, I was fond of plucking

the skinny wings off the craneflies that were so numerous where I grew up in Richmond, British Columbia. Sometimes I'd remove the legs too, leaving the grub-like body squirming in the dirt. This ended when I was around nine or ten, but I've felt guilty about it ever since. Maybe liberating crickets will offset my bad karma.

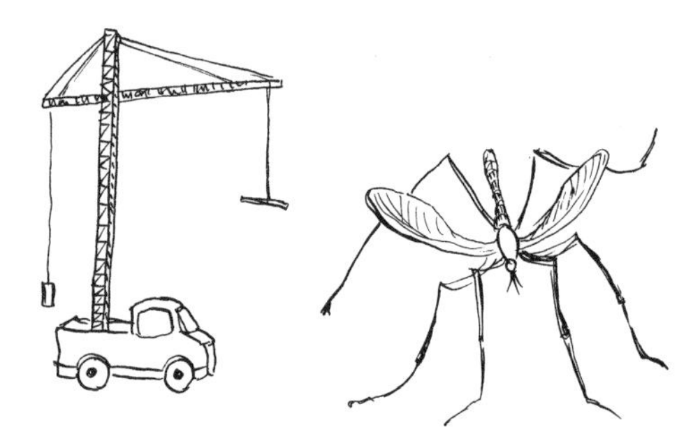

Crane Cranefly

(drawings not to scale)

Male and female craneflies recognize each other only when they happen to bump legs. This touch triggers the beginning of the courtship process. When a male is bumped by a female, he grabs one of her legs and waits for her to signal her okay by raising one of her other legs. This gives him the go-ahead to proceed with the courtship, including kissing the back of her head, and then mating.

Yet another species of fly, fruit flies are a favourite of genetic scientists but not of gardeners. Amazingly, they can produce twenty-five generations in just one year. Swarms of these minute yellow insects are nearly always found near fruit, especially when it's rotting. You'll also find them in your compost bin, helping to dispatch decaying organic matter.

Unfortunately, female fruit flies also like to lay their eggs in fruit that is still growing. The larvae have a readily accessible food supply but the developing fruit is often damaged irreparably.

Aristotle thought fruit flies spontaneously generated from the sediment of sour wine. He also called fruit flies "vinegar flies" and believed they only liked sour things, whereas they are in fact attracted to all kinds of sweet fermenting materials such as decaying fruit.

Nevertheless, humans owe our best genetic research to fruit flies – scientists prefer to call them *Drosophila* – primarily because of their exceedingly short lifespan (two weeks). In 1995 three researchers won the Nobel Prize for their fruit fly research. And I'll bet you didn't know that the *Drosophila melanogaster* genome was mapped in the year 2000? I didn't either, and frankly I don't even know what it means.

Although scientists probably wouldn't agree, I don't find fruit flies to be the most interesting visitors to my green space. But I have to laugh at their cute courtship behaviour: a male courts his desired female by sticking his tongue in her face. Once this attractive ritual is out of the way, the pair dances together with a series of sideways waggles, which culminates in the male mounting the female. And thus the fruit fly's cycle of life begins anew.

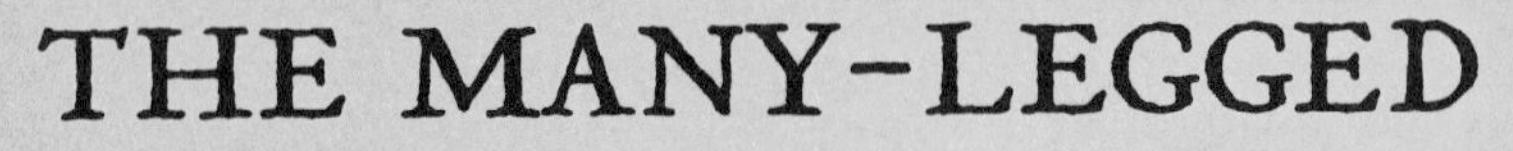

THE MANY-LEGGED

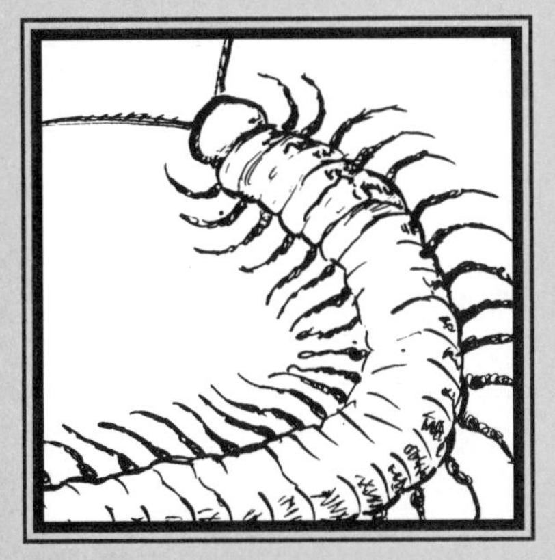

the many-legged

Fifteen years ago in Thailand, I was chased around a tiny one-room hut by a frightening-looking, foot-long creature that turned out to be a centipede. It wasn't actually pursuing me – it was desperately searching for an escape route, and I just kept getting in the way. The experience scared the heck out of me then. But today I realize how blessed I was to have seen such an amazing centipede at such an intimate distance.

The much smaller North American centipedes are common enough in our gardens, but it's rare that we get a chance to look closely at these fast-moving visitors. When we disturb them from their dark hiding places, their multiple legs (as few as thirty) send them quickly scurrying for cover. Like my cat Rinchen, centipedes have longer back legs than front. In the centipede's case, this helps with stability. For poor Rinchen, it means he sways like a bandy-legged sailor. Centipedes move quite fast and, as they do, the length of their stride increases. At full tilt, only a few "feet" will be on the ground at a time.

I feel confident that you'll know one of the 3,000 species of centipede when you see one, even if it isn't the foot-long tropical variety. All have elongated bodies with many distinct body segments, a curious pair of antennae that resemble a droopy Fu-Manchu moustache, and two poisonous, crab-like claws. The centipedes in your garden probably max out at just two and a half inches in length.

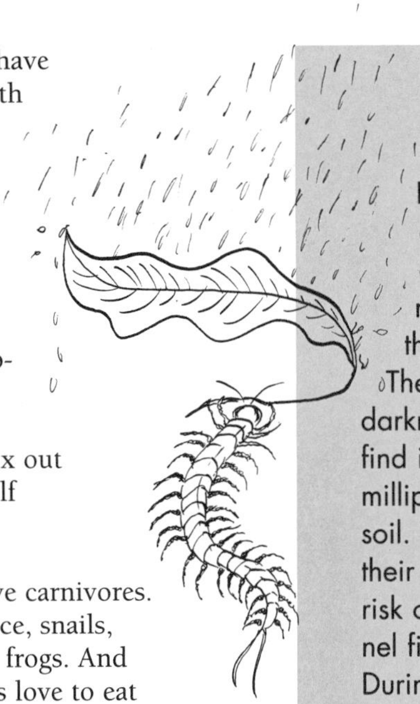

Centipedes are active carnivores. The big guys eat mice, snails, worms, lizards, and frogs. And mice, rats, and birds love to eat centipedes. Farmers love them too because they consume the fly larvae (maggots) that can damage young seedlings. To eat, a centipede waits until prey happens by, and when it does, watch out. The centipede takes charge of the tasty morsel, shooting poison into it through its powerful claws. This sting is painful to animals even of human size.

Exposure to the hot sun can mean death for centipedes and millipedes because they dry out quickly. They prefer the world of darkness. To ensure they find it, centipedes and millipedes burrow into the soil. The big problem with their vertical homes is the risk of drowning if the tunnel fills up with rainwater. During a deluge, you'll find millipedes and centipedes taking cover on your shrubs and other plants. They escape their tunnels and climb up and out of harm's way, using leaves and twigs as shelter.

MILLIPEDES

millipedes

My friends and family know I'm not the world's most coordinated individual. I often lose my balance, trip over my own feet, and sometimes fall over for absolutely no reason. That's why I admire millipedes, who have eighty-eight legs or more. I'm envious of the grace with which they manage all those feet.

The 8,000 millipede species are secretive creatures that are more reclusive than centipedes. They're herbivores, not hunters, so these guests reveal themselves to gardeners infrequently. Like the centipede, a millipede's body is made up of segments. Canadian and American millipedes have at least eleven segments but tropical species can have up to 100. Their legs (two pairs per segment) are better designed for digging than jogging. Millipedes use all their legs at once to burrow into the earth. When its legs push off simultaneously, the force is so great the poor thing must curl its head under its chest to keep it from snapping off.

Millipedes are another of your garden's recycling experts. They're wonderful at breaking down decomposing leaves but are also adept at disposing of your seedlings. Some have a sucking mouthpiece that allows them to pierce your tender plants and suck the sap out of them. Others nibble the roots or eat the entire plant. Even so, Nature has found ways to protect millipedes from pesticide-spraying gardeners. They dig down into the soil, well away from sprays, and their hardened exterior also helps protect them from chemical warfare.

Alas, gardeners are not the gentle millipede's only predator. Spiders, mantids, birds, and lizards also enjoy the taste of millipede flesh. To protect themselves from those who would eat them, millipedes roll up into a tight, tidy ball. In this position the hard plates on their back protect their soft undersides, and their legs tuck conveniently into the spaces between the articulated body segments. If you insist on picking up a millipede, don't put it near your face. They emit a defensive gas from their pores that will sting your eyes.

As you can see, the millipede is prepared

to fend off just about any attempt at eradication. My advice is to leave them be. So what if they eat a few seedlings while they turn your leaf compost into rich organic soil booster? Besides, I'm inclined to think kindly of these slow-moving little pacifists. They do tremendous good in my garden and don't kill other creatures to survive. And as herbivores, they have no need for the scary poison claws of their centipede brothers and sisters.

THE NO-LEGGED

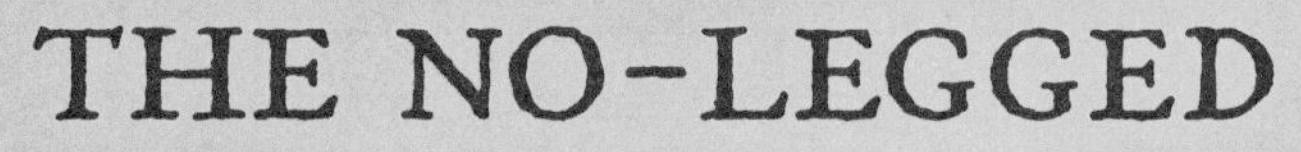

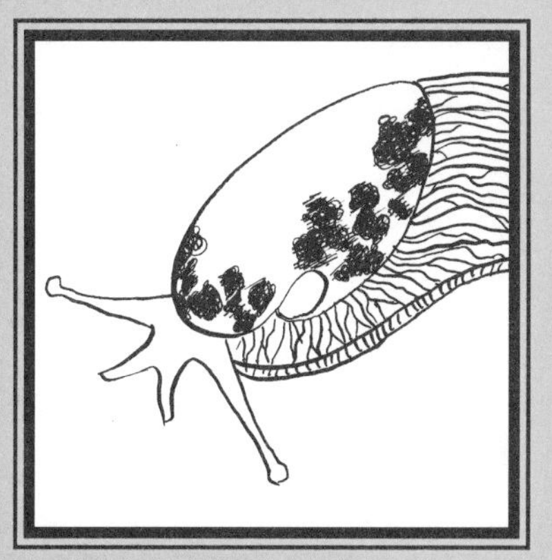

the no-legged

Like most North Americans, you probably salted a slug or two in your youth. What a horrible tradition, given that the entire surface of a slug's body is covered in thousands of highly sensitive nerve endings. No wonder the poor creatures curl up into a fetal position when you poison them in such a manner.

Certainly, humans consider slugs to be one of nature's most despised inhabitants. As a former owner of a garden centre, I can tell you that there's no end to the wicked strategies gardeners conjure up to exterminate slugs. I came to feel very sorry for these slimy creatures when I heard about the barbaric things people do to them: squishing them between boards; pouring poison on their sensitive, vulnerable bodies; drowning them in old margarine containers; asphyxiating them in sealed glass jars. Or the truly terrible practice of collecting slugs in a plastic bag, to be thrown into the street and run over by traffic. And forgive me for lecturing, but how many of you users of slug bait think twice about the harm you do to beneficial guests like the friendly earthworm?

Still with me? Now that I've got that off my chest, let me share some very cool slug facts with you, in the hope of converting you to my team.

A slug's most distinguishing feature is the slick, moist mantle that covers its body. Slugs wear mantles of many colours: green, yellow, black, brown, and spotted too. And as for size, they also run the gamut: big, small, fat, and skinny. Here on the mild, rainy west coast of Canada we have a truly stupendous variety of slugs in our gardens. If you make your home in the equally rainy parts of Washington or Oregon, you're similarly blessed.

Slugs do most of their feeding at night. And feed they do on tender seedlings, newly opened blossoms, and young root systems. It's hard to satisfy these hungry garden guests. But as it feasts, the slug must be mindful of the mice, rats, fireflies, beetles, frogs, ducks, centipedes, and birds who are their natural enemies. So you see, slugs have plenty to worry about without humans plotting to destroy them too.

What are slugs good for? Well, the Germans used to eat them by

A British scientist has found that slugs turn up their antennae at Guinness stout, cider, and some English bitters. Gin-and-tonics and lagers are more popular, but Kaliber, an alcohol-free brew by Guinness, beats them all. Guess there are more than a few teetotaling slugs.

Some slugs have post-coital difficulties. Disengaging is a challenge for two amply endowed banana slugs, both of whom are thoroughly covered in sticky sexual slime. After long bouts of writhing and pulling, the pair may resort to one slug gnawing off the other's penis.

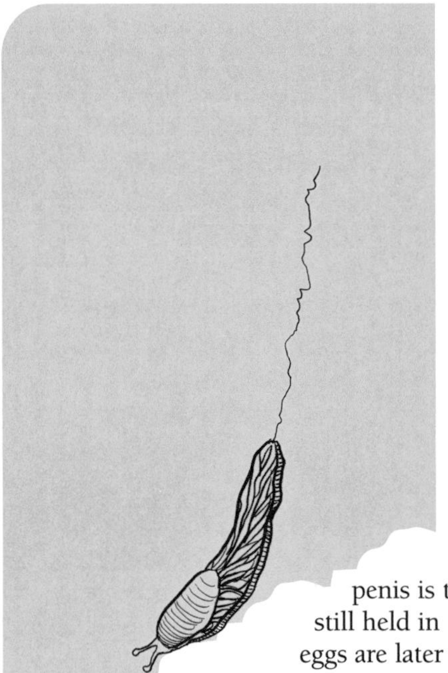

removing the slime with vinegar and then gutting and deep-frying them. And less than ten years ago, scientists explored slug slime as a possible treatment or cure for cystic fibrosis. Slugs also create garden art. Just as I find spider webs shimmering with dew to be beautifully decorative, so too do I find slug trails shining silver in the morning sunlight.

Slugs hold the record for leisurely coitus. As you can imagine, their courtship is a slow process. The two slugs circle each other for several hours, all the while trailing thick sexual mucus. They lick and lap at the gooey stuff and then both eventually unfurl their rather long penises. They twist these startling appendages into a knot for the exchange of sperm. Once this feat is accomplished, each penis is tucked back inside the body and the sperm, still held in the penis, fertilizes the awaiting eggs. These eggs are later laid through pores in the slug's head. (Ouch.)

You can recognize slug eggs quite easily. The squishy, tapioca-sized balls look almost exactly like time-release fertilizer. Slug eggs are usually laid in moist, dark areas, under bits of wood for instance.

As you've no doubt gathered by now, my garden is a no-kill zone. Slugs co-exist with nature's other guests, including my two cats and an army of stray kitties. If you peek over my fence you might see fat cat Rinchen prancing madly around in circles, desperately smacking his tongue against his jowls. Rinchen is loveable but not terribly bright. Unlike Tsewang, he's never learned to not eat a slug.

We humans often laugh at the slow pace of slugs and snails. Telling someone they "move at a snail's pace" is rarely a compliment, given that these guests move at top speeds of 0.025 mph to 0.03 mph depending on the species.

The shell is what separates the snails from the slugs. Interestingly, the shell nearly always hangs to the right. Left-oriented shells are considered collector's items by snail aficionados. The shell is made of a thin layer of protein laid over several hard layers of calcium carbonate – the same mineral women take to prevent osteoporosis. When a snail disappears inside its shell, it goes head first to protect its most sensitive bits. The foot is tucked in last. When retreat is complete, the shell opening is closed off with a flat, hardened part of the foot.

The garden snail takes shelter inside its shell by day, and emerges at night to feed, although I recently noticed half a dozen snails at varying heights in my fledgling lilac tree – in broad daylight. Maybe they were confused by the dull winter sun.

Some snails are carnivorous, while others are vegetarians. Speaking of carnivores,

"Worm nor snail do no offence." – *A Midsummer Night's Dream*

the French relish an appetizer of *Helix pomatia*, but will eat the more common garden snail *Helix aspera* when *pomatia* supplies run short. Napoleon ensured his soldiers didn't go without: he supplied 1,000 snails per man per week during his campaigns. These were considered to be "emergency" rations.

Like slugs, snails are hermaphrodites: they have both sexual organs, allowing them to mate with whoever comes along. How handy! The round, pale eggs are deposited on leaf mould and other organic debris in your garden. When they hatch, the newly emerged snails have no shell to speak of so the little creature attaches itself some place safe, like a tree, until one grows. In winter, snails hibernate. You might see their shells, sealed up, in sheltered spots in your garden.

Amadea, my comrade in cricket release, is also known as Amadea the Snail Charmer. Instead of capturing garden snails to eat or destroy, this three-year-old encourages them to affix to her arms, legs, fingers, and toes. Amadea the Snail Charmer can be overheard gently cooing and holding one-sided (we think) conversations with the creeping mollusks. She gives each a name and they actually appear to respond to her charming efforts.

I recently learned of a trick my young snail girl might enjoy. Placing a snail or slug on a piece of glass provides a clear view of a snail's mouth. Apparently you can see it scrape at food with its rough, file-like tongue.

Snails can lift ten times their own weight up a vertical surface. This is like a person lifting a ton.

Thomas Greene of La Plata, Maryland consumed 350 edible snails in just eight minutes, twenty-nine seconds. Perhaps Mr. Green has French blood: the people of France eat nearly 50,000 tonnes of snails every year.

Earthworms are the perfect guests. They're cute, clean, tidy, and your garden won't thrive without them. Worms live on and in the decaying organic matter in your soil. Your compost heap is probably their favourite hangout. Some gardeners buy red-wiggler worms to boost decomposition. How exactly do earth-worms bring such balance to your garden? Much of an earthworm's time is spent excavating burrows. As they dig, they contribute to the good health of your garden by aerating the soil.

Worms swallow soil as they dig their burrows and poop it out alongside the entrance. Polite people call earthworm poop "black gold" or "worm casings." I say poop is poop. You can purchase it at your local garden centre – it's one of the best organic soil amendments you can buy.

The body of an earthworm is made up of many segments, each one exactly like the others, except for the head and the rear end. The average-sized North American earthworm has about 150 segments spread over its ten-inch length. Worms breathe through their skin so the surface must stay moist at all times. If it dries out, the diffusion of oxygen throughout the body crashes to a halt and the earthworm dies an unpleasant death. To avoid heat and sunlight, earthworms are primarily nocturnal. Those worms you see wriggling across your concrete patio have either stayed out too late, or lost their way. Do these beneficial guests a favour – when you see one baking on hot, dry pavement, take a moment to gently carry it to soil or grass. That's what the Tibetan Buddhist lamas do.

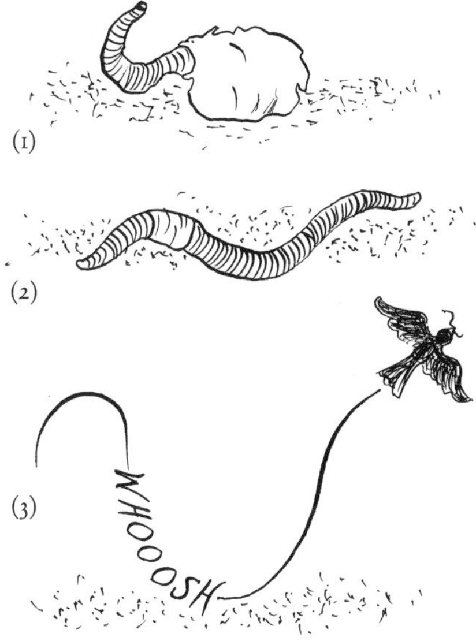

Life Stages of the Earthworm

Earthworms were once considered to be good for damaged sinews, mending broken bones, ear conditions, toothaches, wounds, nosebleeds, various ulcers, removal of corns, gout, jaundice, tertian fevers, asthma, breast conditions, stimulating breast milk, promoting afterbirth, and as an antidote to the scorpion's sting.

The classical Greeks believed earthworms spontaneously generated in wet soil. Some also subscribed to the theory that earthworms turned into eels.

Consider the poor earthworm next time you dose your garden with chemicals like slug bait because your irises are being eaten. As the earthworm eats your soil and garden refuse, it also ingests whatever toxins you've introduced. These accumulate in its body in highly concentrated amounts until it eventually dies. Robins and other birds snack on earthworms and they too can die depending on the level of chemicals in worms they've eaten.

Earthworms won't live in soil that is overly acidic or alkaline. If you have few earthworms in your garden, try a soil test. You may find that the soil is outside the worm's comfort range of pH 4.2 to 4.5. And remember that pesticides and other chemicals can dramatically alter the pH of your soil.

Leeches aren't guests in all North American gardens. They're found mostly in warm, humid places and in or near ponds, streams, lakes, and rivers. But because the little slimers also live on land in cool, damp areas, don't think you're completely safe from a leech visitation. Leeches are found even in Antarctica.

While leeches can grow to a whopping eighteen inches, most are only an inch or two long. Primarily black or brown, leeches can also be striped or spotted. They are keenly aware of vibration, which usually indicates a food source is nearby. Their entire body is covered in special receptors that are triggered by movement in the air, land, or water around them. When the receptors are fired, the leech stands straight up and then, like a gymnast, flips end over end towards its prey.

To attach to the host, the leech uses teeth or a piercing proboscis (like a mosquito's) to puncture the host's skin and draw blood. A leech bite is painless thanks to numbing chemicals in leech saliva. I de-leeched many of my hiking buddies in Nepal years ago – no one ever felt a thing until I tried to detach the blood-filled critter!

In Napoleon's time, the demand for leeches was so high that the medicinal leech became an endangered species. Laws were passed to regulate the sale of leeches, and breeding farms were established to restore the leech population.

Leeches allowed to rot in vinegar or their ashes applied with vinegar were believed to be effective both for the removal of unwanted hair and also as hair dye.

Leeches have been used medicinally for thousands of years. Early Greeks thought that leeches could bleed the presence of evil spirits out of the afflicted person's body. Later, leeches were used to combat a variety of illnesses, most of which were thought to be caused by an inflamed digestive tract. Blood-letting using leeches was considered the most effective way to bring down swelling.

Although the craze for leeching dissipated in the late nineteenth century, leeches are still used medicinally today. Many people swear that application of a leech is the most effective way to cure severe bruising or a black eye. And some doctors use leeches to decrease swelling and pain when body parts are reattached.

So while leeches may be among the creepiest guests in our gardens, they have many redeeming qualities. You might want to follow the advice of the ancient writer Marcellus, who believed that a live leech kept in a nutshell and hung around the neck as an amulet was effective against conditions of the palate.

BUGS: THE REAL THING

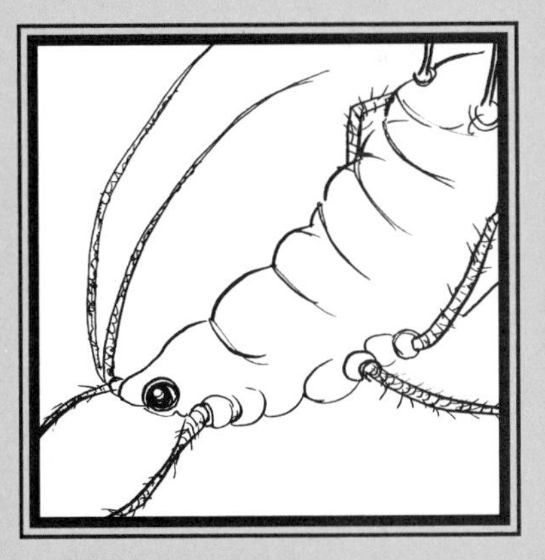

bugs: the real thing

Bugs: The Real Thing

You thought this whole book was about bugs, didn't you? Well, it is. Sort of. When gardeners say "bugs" we're really using the term as a catch phrase for all crawling, flying, creeping, sliming creatures. But true bugs are their own entity. And they've waited a long time for recognition. Unfortunately, I suspect most of us will always call an insect a bug and a bug a bug.

All bugs have a beak-like shnozz with a piercing and sucking mouth. This is what distinguishes bugs from other insects. Bugs use this handy sucking device to extract sap from plants and sometimes blood from animals and humans.

There are two distinct groups within the bug family: carnivorous waterbugs, and herbivores like aphids, scales, spittlebugs, cicadas, mealybugs, and sowbugs.

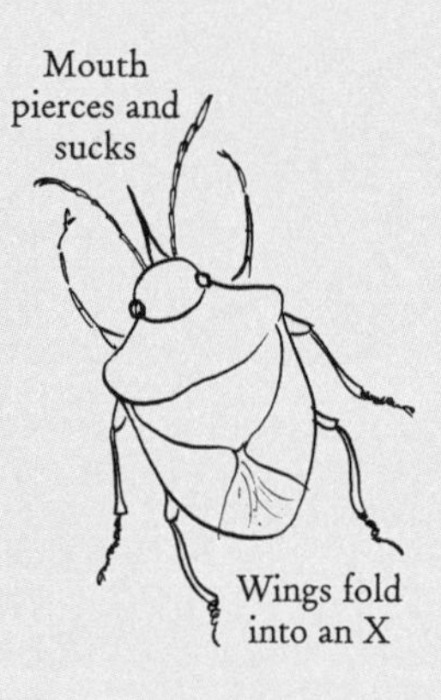

In Mexico, waterbugs are called "Jesus bugs" for their ability to walk on water. But unlike the holy man, waterbugs are predators.

Carnivorous waterbugs float on the surface of water sources in your garden, waiting to snack on insects unfortunate enough to fall into the drink. As these insects struggle for their lives, they disturb the water's surface and signal to the waiting waterbug that dinner has arrived. Even the slightest shiver sends out ripples that are detected by the 25,000 species of waterbugs like waterstriders, waterskippers, and pondskaters.

Waterbugs use their beak to catch their prey. When it's not in use, the beak is kept tucked under the waterbug's body, and swings out when a likely suspect approaches. Waterbugs also have large eyes to help identify prey and detect danger.

Waterbugs have developed several clever ways to avoid being eaten by other bugs and insects. Many use stink glands to repel would-be killers. Waterstriders, the rowing champs of the bug world, use their long, gangly legs as oars to propel themselves away from predators. They can

do this even when clutching dinner in their short but strong fore legs.

Some waterbugs live below the water's surface, eating snails, fish, and insects. They have a strawlike "tail" that connects them continually to the oxygen supply above. The need for oxygen means underwater bugs don't do a whole lot of swimming. Mostly they hang around just below the surface, keeping deathly still if a food source is near, and then reaching out with pincered fore legs if it comes within striking distance.

I like to think of the 4,000 species of herbaceous aphids as miniature landscape architects. The curly, twisted deformation of plant material caused by aphid sap sucking lends slightly surreal structure to the greenery in my garden.

Aphids are very small bugs (less than one-eighth of an inch) that form very large colonies. They come in a variety of shades: red, black, pink, white, or green. Like all true bugs they have a long beak with a piercing, sucking stylet.

In one respect, they're a rarity among the guests in your garden – they're one of the only bugs and insects to have live births. That means larvae are ready to begin drinking sap from the leaves and roots of tender young plants right from the get-go. Their voracious feeding causes the wilting, gall, weakening, or death of our favourite plants.

At my gardening centre, the number one insect question people asked was: "How do I get rid of aphids?" Most gardeners found little consolation in our suggestions to use organic or biological controls, or to simply do nothing. Like slugs, aphids seem to fuel a "get out or be exterm-

In the Middle East, people eat a sweet made from the honeydew of aphids. The Kurds of Turkey and northern Iraq collect large quantities of honeydew from aphid-infested oaks. Branches are cut in the early morning hours (to gain a head start on honeydew-loving ants) and then shaken vigorously to knock off the honeydew. The sticky stuff hardens quickly in the dry desert air and the "rocks" of honeydew can then be dissolved in water and mixed with eggs, almonds, and seasonings. This confection is boiled and allowed to resolidify, whereupon it's cut into pieces and sprinkled with sugar. I'm told it's absolutely scrumptious.

inated" reaction in otherwise compassionate green thumbs.

How about learning which plants attract aphids in your plot of earth? Then you can avoid these plants entirely, or plant them for the purpose of luring the aphids away from your more prized specimens. I can personally guarantee that nasturtiums rapidly turn into aphid condos. Plant them well away from your favourite greenery.

When aphids feed, they produce copious amounts of feces. This poop is what many gardeners quaintly call "honeydew" or "Indian sugar." It is a sticky, rather annoying residue that is stubbornly resistant to removal from patio tables and chairs, and fans the fires of anti-aphid sentiment.

Honeydew attracts ants, who are great friends of aphids. The ants want a steady supply of honeydew. The aphids need protection, as well as a shuttle service to and from juicy plant material around your garden. Yet another wonderful example of cooperation in the insect world.

Scale are another small bug that form large colonies and, like their aphid cousins, can turn your plants into striking garden sculptures. Too bad most gardeners are unappreciative of the efforts of these hard-working vegetarians – not only do they produce honeydew, but they make shellac too. And they were once used to "cure" dysentery and eye problems.

The scale world, 6,000 species strong, is run by females – nearly all adult scale are female. It seems that the sole purpose of what few male scale there are is to help the females to reproduce. These gal scale annoy some gardeners with their feeding and pooping and honeydew production. The female mates and then dies after laying up to 3,000 eggs under her hard exterior shell. Even unfertilized females lay eggs, which explains scale's ability to rapidly expand their population when they settle on a nice place to live. Look for their attractive shells on the limbs and branches of your fruit and nut trees.

Scale come in a variety of shapes and sizes. My personal favourites are the

oyster and mussel varieties that are shaped like the sea creatures. The resident female scale lives in the far end of her shell, saving the rest of the space for her eggs.

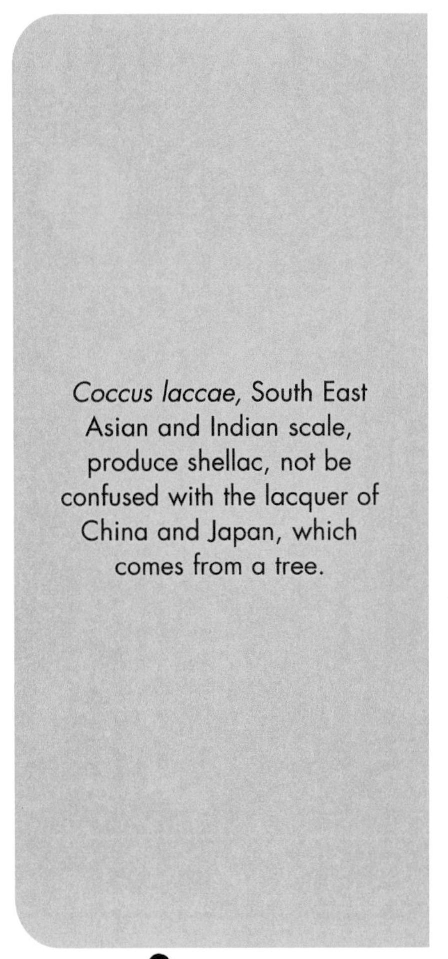

Coccus laccae, South East Asian and Indian scale, produce shellac, not be confused with the lacquer of China and Japan, which comes from a tree.

Cottony Cushiony Scale

Oyster Shell Scale

San Jose Scale

Bathroom Scale

Sap-sucking spittlebugs feed on the leaves in your garden. The undersides of leaves and stems of tall grass often contain the spitball of these pale green bugs. The frothy mess protects these tiny beings from rain, heat, and ants. Adults look like miniature frogs, hence the name "frog-hopper." Even when fully mature, a spittle-bug is less than a third of an inch long.

I'm sure the spittlebug wishes its spit would protect it from small children too. As a child, I must have destroyed the homes of a hundred spitbugs. I loved squishing that soft froth between my fingers to find the little green guy, who I then obliterated without a speck of remorse. It was endlessly entertaining, but accumulated some very negative karma.

To make the spit, the spittlebug uses the stylet on its beak to bore into the stem of a plant and suck out the sap, which is then expelled from the creature's throat with enough force to froth it – like a mini cappuccino machine. The spitbug loads the foam onto its back with its hind legs, eventually concealing its entire body within the goo.

Cicadas are large plant-eating bugs that can reach nearly three inches in length. In North America, we're fortunate to have over 3,000 species of these singing herbivores. It's a bit difficult to explain what the winged, six-legged cicada looks like. I think they resemble a cross between a bee, a grasshopper, and a beetle grub. But unlike the grasshoppers that people sometimes confuse them with, cicadas do not jump. They begin to appear each year in midsummer, and live in round holes in the ground about as thick as your thumb and about twenty inches deep.

You'll know a cicada is close by when you hear a series of unusually loud clicks that can be heard by humans half a mile away. This is the sound of a male cicada courting a female. He "sings" by using sound-producing equipment in his belly. Two membranes on either side of the belly are contracted and then relaxed using powerful abdom-inal muscles. As the membrane moves, it buckles and returns to its original shape over and over. The buckling is the source of the clicks. Cicadas also have air sacs to amplify the sounds.

Cicadas are long-lived – some grow into their teens – but most of their lifespan is spent as larvae. The larvae live underground, where they suck the liquid out of plant roots. It can take more than ten years for larvae to grow to adulthood, due largely to the poor nutrition provided by root sap. When they eventually climb out of the ground and into the trees to molt into adults, the new cicadas must be leery of birds and other predators until the outer cuticle of the body hardens and its camouflage colours develop. You may have seen the skin outgrown by the nymphs and left behind on trees in the early summer.

Gardeners residing in the northern part of the United States will have seen the extensive damage wrought by *Magicicada septendecim*. Their eggs are laid in the trees of an orchard or forest. In several weeks, when the eggs hatch, the larvae (so tiny you can just barely see them) fall to the earth on silken threads, where they dig in for an extended period of debauchery. During this time the once green trees turn brown because under the ground, larvae are sucking the plant roots dry.

"They were singing on this planet before us; they will sing after us, celebrating what can never change, the fiery glory of the sun."
Jean Henri Fabre, *The Wonders of Instinct*

Aristotle, in *History of Animals,* wrote that cicada nymphs taste best when they first come out of the ground, before they shed their skins. He also admired the taste of adult male cicadas, and of females big with eggs.

Closely related to scale, mealybug females have no wings, while the midget males are blessed with two. But unlike scale, mealybugs can move about, albeit slowly. They make their homes on the undersides of plant leaves, gaining nourishment by sucking the sap out of the plant (and thereby stunting the plant's growth).

Although rarely safe from gardeners, these creatures safeguard their lives from other insects by secreting and surrounding themselves with a waxy, whitish coating. They live in colonies so numerous that a group of mealybugs looks like a tuft of cottonwool.

Mealybugs join forces with aphids in their mission to cover the universe with honeydew. Or, maybe it's just their way of being nice to ants.

Sowbugs (a.k.a. pillbugs or woodlice) are sometimes confused with mealybugs. But these common residents of your garden are actually a type of crustacean – the same family as shrimp. They enjoy visiting your garden if you provide damp places for them to hide out during the day, and plant debris for them to break down for food at night. One of the great things about having sowbugs as guests is their superior ability to help you compost. They do much good in our gardens, feeding on decaying organic matter like compost and leaf piles. They turn dead plants into usable organic matter lickety-split.

Sow Sowbug
(drawings not to scale)

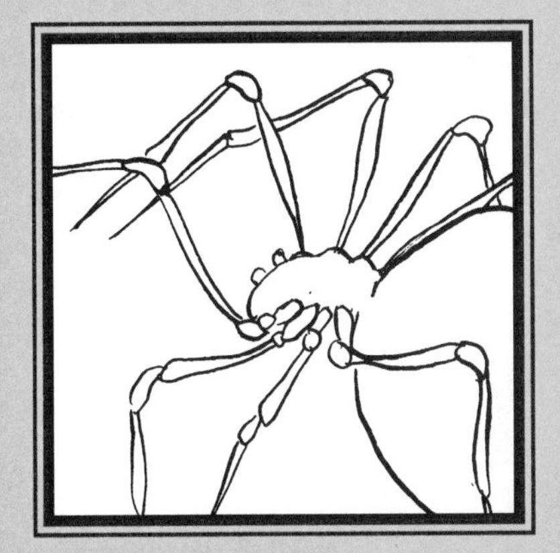

the not-so-scary after all

Spiders are the least welcome guests in my garden. I admit to being irrationally scared of them. At this very moment in my green space there are thousands of spiders. And it's less than 200 square feet!

Spiders inhabit a variety of places – they live in the trees and shrubs, under leaves, on the soil surface, and tunnelled below it too. There are nocturnal spiders and spiders active during daylight. Living where I do in Vancouver, the chance of a poisonous spider taking up residence in my garden is pretty darn remote; only a small minority of spiders anywhere in the world are poisonous.

There are over 3,000 spider species in North America. These members of the Arachnid family have no wings or antennae. Instead, nature blessed them with two mouths, eight hairy legs, and two rows of four eyes. In some species, like wolf spiders, two of these eyes are greatly enlarged.

Spiders perform a wonderful service in our gardens; they catch and eat large quantities of other insects, so many, in fact, they could officially be designated as population controllers. That's why my fear of spiders

doesn't translate into killing them. When one makes its way into my house, I gingerly place a mason jar over it and slide a piece of paper under the spider's feet. The object of my fear is safely contained for the trip out to the garden, where it's released and put to work catching insects.

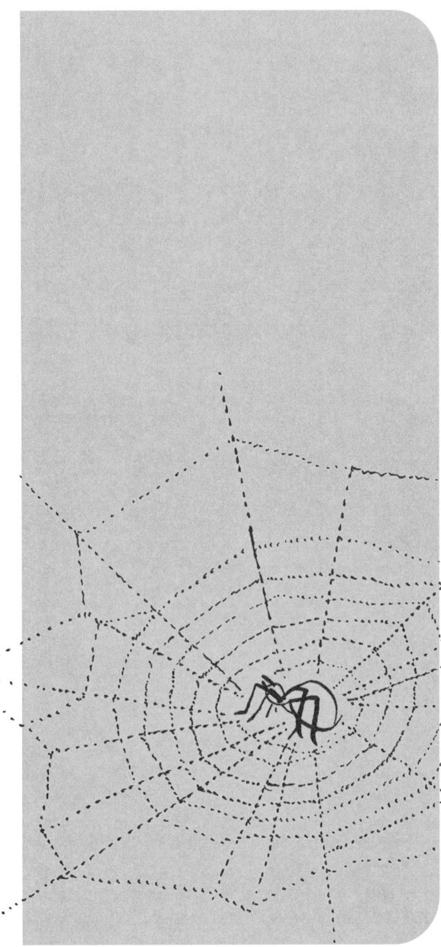

Only ten percent of spiders are capable of spinning the webs we so admire in our gardens. A web provides its owner with a place to hang out, and a means to catch its food – a single web can catch hundreds of insects in just one day. Most web spinners create an entirely new web each dawn; in this amazing process, the spider takes its old web, rolls it up, dissolves it with saliva, and then drinks it.

Nature never fails to humble and amaze me. She's given spiders the ability to produce different weights and types of silk, depending on what they need it for. Silk is used to wrap prey, to drag possessions, to build nests, as safety nets and egg sacs, and yes, for spinning webs.

As with most garden guests, spiders have

their own special way of getting it on. When he's feeling amorous, a male spider may court a female by playing her web like a harp. To me, the most amusing courtship ritual of all is performed by the male nursery-web spider, who presents a nicely silk-wrapped package to the female in the hopes of winning her over. This could be a bee, fly, beetle – or, if he's tricky, a rock.

A female spider lays tens of thousands of eggs during her lifetime. The way they nurture their eggs differs between spider species. A mama wolf spider carries her eggs in a sac attached by webs to her belly. Pity the poor wolf spider – its sac is as big as she is. A nursery-web female does the same thing, but the egg sac attaches to her fangs! And it's so large that she has to walk around on her tip-toes. Who could blame the many other spiders that simply abandon their eggs?

Wolf spiders

There are over 100 species of wolf spiders in North American gardens. They're usually grey or brown, and have two very large eyes and six smaller ones. Wolf spiders are wanderers and rarely keep permanent homes. Oh, and they're hairy and really big. I once saw one move a bathmat-sized piece of cloth thrown over top of it!

Orb-weaver spiders
Often black and yellow. There are hundreds of species in Canada and the U.S. Orb-web spiders spin the pretty webs we see in our gardens.

Huntsmen spiders
Found in the southern United States – and in boxes of fruit throughout Canada and the rest of the U.S. They resemble tarantulas.

Jumping spiders
Like wolf spiders, the 300 North American species of jumping spiders don't maintain permanent shelters. If you see a brightly coloured spider, small and squat, jumping here and there in your garden – it's probably a jumping spider.

Tarantulas
There are more than thirty species in the U.S., mostly in the arid southwest. Tarantulas live longer than most other spider species. These extremely hairy guests can be nearly as old as I am (thirty-eight, in case you're wondering). Look for silk-lined nests in the ground or in cracks in rock faces. Several species make their nests in trees, where they catch birds!

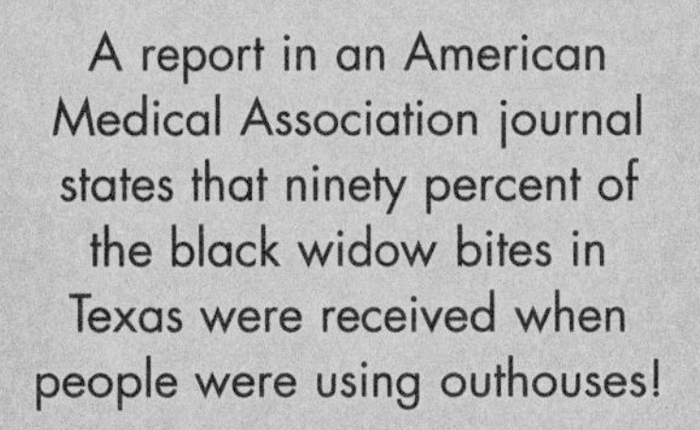

A report in an American Medical Association journal states that ninety percent of the black widow bites in Texas were received when people were using outhouses!

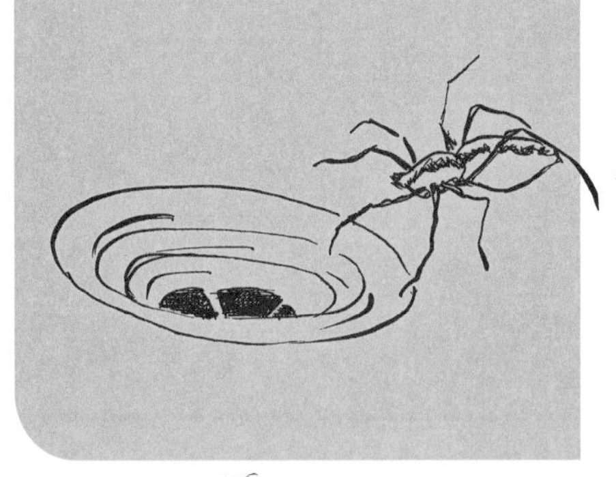

Comb-footed spiders
Among the comb-foot spider species is the infamous black widow spider. If you live in the southern United States (or in some northern states), you may have seen a black widow. They have a humped abdomen and spin untidy webs.

Six-eyed spiders
Don't get too close. The species includes the brown recluse spider, found in the south and central United States. Easily recognized by the violin-shaped marking on its thorax, the recluse likes to hang out in gardens, closets, and basements. Its bite can become a nasty ulcer if left untreated.

Nursery-web spiders
These are larger versions of wolf spiders. The female nursery spider makes a little web tent in a bush or tree or crevass to contain her egg sac. She carries the egg sac in her mouth, places it securely within the tent, and then stands guard at the entrance. They're adaptable creatures and can survive in semi-aquatic environments. They're big, they're hairy, and – although I'm told it's just not possible – I'm convinced a family of them lives in my bathtub plumbing.

Another member of the Arachnid family, the scorpion has large pincers that give it an assertive appearance, and make confusing one with anything but a midget lobster nearly impossible.

Scorpions are nocturnal. They rest by day under rocks, bark, and logs; at night they come out to feast on sleeping day-active insects, mice, and even small lizards. Like spiders, they help keep the guests in the garden list down to manageable levels.

I thought these confident creatures lived only in hot, dry parts of the United States like Arizona. But no, the 1,200 species of North American scorpions have been found living as far north as the desert areas of British Columbia's Okanagan. They can even be found in B.C.'s damp rainforests. Scorpions can survive long periods of drought and hibernate if it gets too cold.

The scorpion's stinger is located at the tip of its slender segmented tail, which is held over its body. Depending on the species of scorpion, the venom ejected from the stinger can be painful but non-toxic, or it can be lethal. Gardeners

working in arid, rocky spots should be mindful when moving stones. Keep in mind that they can grow as large as seven inches.

Scorpion babies, once hatched from their eggs, get a free ride from their mother, riding on her back until they have developed into young adults. At this point, Mum boots them off to fend for themselves, which they do quite nicely.

Mites and ticks are also members of the Arachnid family. They're red, black, brown, or yellow spider-like beings with eight legs and absolutely no division between the head, thorax, and belly. So far scientists have named 100,000 mites and 30,000 ticks.

Mites are common guests in virtually all North American gardens. They're *usually* beneficial too – eating many of the plant-eating critters in your garden. Look for them in leaf litter, soil, bark, or grass. But they also enjoy sucking the sap out of your plants if there are not enough insects on the menu. Some mites, like spider mites (red and less than three one-hundredths of an inch long), are efficient travellers. Unfortunately their journeys can spread plant fungi and viruses from one area of your garden to another. You can prevent this by cleaning up contaminated debris, and by maintaining healthy, happy plants.

Ticks are larger (up to half an inch long) and are much more dangerous than mites. For all the fear that surrounds spiders, it is the tick that is the most harmful arachnid to humans. Among the

Europeans put a type of mite into cheese. The mite introduces the mould that gives the distinct taste, stinky smell, and veined appearance.

potentially fatal diseases ticks can transmit are Rocky Mountain Spotted Fever, Lyme Disease, and Colorado Tick Fever.

Ticks have highly refined sense organs in their front legs that tell them when a human or other animal is approaching. They climb as high as they can on whatever plant is nearby and leap onto their new home. Using tiny, serrated pincers, the tick pierces the skin of its host and anchors in to suck some blood. Some ticks drop off after feeding, while others burrow into the host's flesh, transmitting disease as they dig. At least ticks won't eat your flowers!

BIBLIOGRAPHY

Beavis, Ian C. Insects & Other Invertebrates in *Classical Antiquity*. Exeter: University of Exeter, 1988.

Berenbaum, May R. *Buzzwords: A Scientist Muses on Sex, Bugs, and Rock'n'Roll.* Washington: Joseph Henry, 2000.

Bodenheimer, F. S. *Insects as Human Food.* The Hague: Dr. W. Junk, Publishers, 1951.

Chinery, Michael. *Spiders*. London: Whittet Books, 1993.

Clausen, Lucy W. *Insect Fact and Folklore*. New York: Macmillan, 1954.

Corbel, Eve. *The Little Greenish-Brown Book of Slugs*. Vancouver: Arsenal Pulp Press, 1993.

Darling, Lois and Louis. *Worms*. New York: William Morrow, 1972.

Darwin, Charles. *The Formation of Vegetable Mould through the Action of Worms with Observations on their Habits*. London: Faber and Faber, 1945.

Davy, Alfred. *Gilly: A Flyfisher's Guide*. Vancouver: BC Flyfishers, 1981.

Fabre, Jean Henri. *Insects*. New York: Gluck Press, 1979.

Gordon, David George. *Field Guide to the Slug*. Seattle: Sasquatch Books, 1994.

Harper, Alice Bryant. *The Banana Slug*. Aptos, CA: Bay Leaves Press, 1988.

Henwood, Chris. *Keeping Minibeasts: Snails and Slugs*. London: Franklin Watts, 1988.

Hornblow, Leonora and Arthur. *Insects do the Strangest Things*. New York: Step-up Books, Random House, 1990.

Hoyt, Erich and Ted Schultz (eds.). *Insect Lives: Stories of Mystery and Romance from a Hidden World*. New York: John Wiley &Sons, 1999.

Imes, Rick. *Incredible Bugs*. Toronto: Macmillan-Quarto, 1997.

Jennings, Terry. *Junior Science: Earthworms*. East Grinstead, Sussex: BLA Publishing, 1988.

Kneidel, Sally. *Pet Bugs*. New York: John Wiley & Sons, 1994.

Lauck, Joanne Elizabeth. *The Voice of the Infinite in the Small*. Mill Spring, NC: Swan·Raven-Blue Water, 1998.

Laughlin, Robin Kittrell. *Backyard Bugs*. San Francisco: Chronicle Books, 1996.

O'Hagan, Caroline. *It's easy to have a snail visit you*. London: Culford Books, 1980.

Olney, Ross and Pat. *Keeping Insects as Pets*. New York: Franklin Watts, 1978.

O'Toole, Christopher. *Alien Empire*. London: BBC Books, 1995.

O'Toole, Christopher. *The Encyclopaedia of Insects*. London: Andromeda Oxford Limited, 1986.

Patterson, Robert. *The Natural History of the Insects Mentioned in Shakespeare's Plays*. London: A. K. Newman, 1842.

Pearse and Buchsbaum. *Living Invertebrates*. Pacific Grove: The Boxwood Press, 1987.

Pleasant, Barbara. *The Gardener's Bug Book*. Vancouver: Whitecap, 1994.

Preston-Mafham, Rod and Ken. *The Natural History of Insects*. Ramsbury, England: Crowood Press, 1996.

Putnam, Patti and Milt. *North America's Favorite Butterflies*. Minocqua, Wisconsin: Willow Creek Press, 1997.

Ritchie, Carson I. A.. *Insects, the Creeping Conquerors*. New York: Elsevier/Nelson, 1979.

Roach, Mary. "The Instructress" in *The Adventure of Food*. Sterling, Richard. ed. San Francisco: Travellers' Tales, 1999.

Roberts, Hortense Roberta. *You Can Make an Insect Zoo*. Chicago: Children's Press, 1974.

Souza, D. M. *Insects in the Garden*. Minneapolis: Carolrhoda Books, 1991.

Stevens, Carla. *Insect Pets: Catching and Caring for Them*. New York: Greenwillow, 1978.

Sutton, Mark Q. *Insects as Food: Aboriginal Entomophagy in the Great Basin*. Menlo Park, CA: Ballena Press, 1988.

Taylor, Ronald L. *Butterflies in My Stomach*. Santa Barbara: Woodbridge Press, 1975.

Villiard, Paul. *Insects as Pets*. Garden City, NY: Doubleday, 1973.

Waldbauer, Gilbert. *The Handy Bug Answer Book*. Detroit: Visible Ink, 1998.

Watts, Barrie. *Keeping Minibeasts: Beetles*. London: Franklin Watts, 1989.

Westcott, Cynthia. *The Gardener's Bug Book*. Garden City, NY: Doubleday, 1964.

Wheeler, William Morton. *The Fungus-Growing Ants of North America*. New York: Dover, 1973 (1907).

MICHELE DAVIDSON is one of three friends who started Figaro's Garden, Vancouver's funkiest neighbourhood garden centre. Today she works as a fundraising consultant to large Canadian and U.S. environmental NGOs. Michele is also the editor of *Exhibitions: Tales of Sex in the City* (Arsenal Pulp Press, 2000).

EVE CORBEL is a writer, editor, and comics artist. She wrote and illustrated the book *Don't Say No, Just Let Go* (Arsenal Pulp Press), published under the title *Power Parenting Your Teenager* in the U.S. She is also the author of *The Little Greenish-Brown Book of Slugs,* and the comics editor for *Geist* magazine. She lives in Vancouver.